D1217673

ASTRONOMY AND ASTROPHYSICS LIBRARY

Eugene F. Milone
and William J. F. Wilson

Solar System Astrophysics

Background Science and the Inner Solar System

Eugene F. Milone
Dept. Physics & Astronomy
University of Calgary, Calgary
2500 University Drive NW
Calgary, Alberta T2N 1N4
Canada
milone@ucalgary.ca

Willam J. F. Wilson
Dept. Physics & Astronomy
University of Calgary, Calgary
2500 University Drive NW
Calgary, Alberta T2N 1N4
Canada
wjfwilso@ucalgary.ca

Cover Illustration: "Mysterium Cosmographicum" by David Mouritsen

Library of Congress Control Number: 2006940069

Astronomy and Astrophysics Library Series ISSN 0941-7834

ISBN: 978-0-387-73154-4
ISBN: 978-0-387-73153-7 (set)
e-ISBN: 978-0-387-73155-1

Printed on acid-free paper.

9 8 7 6 5 4 3 2 1

springer.com

Preface

This work is appearing in two parts because its mass is the result of combining detailed exposition and recent scholarship. Book I, dealing mainly with the inner solar system, and Book II, mainly on the outer solar system, represent the combined, annually updated, course notes of E. F. Milone and W. J. F. Wilson for the undergraduate course in solar system astrophysics that has been taught as part of the Astrophysics Program at the University of Calgary since the 1970s. The course, and so the book, assumes an initial course in astronomy and first-year courses in mathematics and physics. The relevant concepts of mathematics, geology, and chemistry that are required for the course are introduced within the text itself.

Solar System Astrophysics is intended for use by second- and third-year astrophysics majors, but other science students have also found the course notes rewarding. We therefore expect that students and instructors from other disciplines will also find the text a useful treatment. Finally, we think the work will be a suitable resource for amateurs with some background in science or mathematics. Most of the mathematical formulae presented in the text are derived in logical sequences. This makes for large numbers of equations, but it also makes for relatively clear derivations. The derivations are found mainly in Chapters 2–6 in the first volume, *Background Science and the Inner Solar System,* and in Chapters 10 and 11 in the second volume, *Planetary Atmospheres and the Outer Solar System.* Equations are found in the other chapters as well but these contain more expository material and recent scholarship than some of the earlier chapters. Thus, Chapters 8 and 9, and 12–16 contain some useful derivations, but also much imagery and results of modern studies.

The first volume starts with a description of historical perceptions of the solar system and universe, in narrowing perspective over the centuries, reflecting the history (until the present century, when extra-solar planets again have begun to broaden our focus). The second chapter treats the basic concepts in the geometry of the circle and of the sphere, reviewing and extending material from introductory astronomy courses, such as spherical coordinate transformations. The third chapter then reviews basic mechanics and two-body systems, orbital description, and the computations of ephemerides, then progresses to the restricted three-body and n-body cases, and concludes with a discussion of perturbations. The fourth chapter treats the core of the solar system, the Sun, and is not a bad introduction to solar or stellar astrophysics; the place of the Sun in the galaxy and in the context of other stars is described, and radiative transport, optical depth, and limb-darkening are introduced. In Chapter 5, the structure and composition of the Earth are discussed, the Adams–Williamson equation is derived, and its use for determining the march of pressure and density with radius described. In Chapter 6, the thermal structure and energy transport through the Earth are treated, and in this chapter the basic ideas of thermodynamics are put to use. Extending the discussion of the Earth's interior, Chapter 7 describes the rocks and minerals in the Earth and their crystalline structure. Chapter 8 treats the Moon, its structures, and its origins, making use of the developments of the preceding chapters. In Chapter 9, the surfaces of the other terrestrial planets are described, beginning with Mercury. In each of the three sections of this chapter, a brief historical discussion is followed by descriptions of modern ground-based and space mission results, with some of the spectacular imagery of Venus and Mars. The chapter concludes with a description of the evidence for water and surface modification on Mars. This concludes the discussion of the inner solar system.

The second volume begins in Chapter 10 with an extensive treatment of the physics and chemistry of the atmosphere and ionosphere of the Earth and an introduction to meteorology, and this discussion is extended to the atmospheres of Venus and Mars. Chapter 11 treats the magnetospheres of these planets, after a brief exposition of electromagnetic theory. In Chapter 12, we begin to treat the outer solar system, beginning with the gas giants. The structure, composition, and particle environments around these planets are discussed, and this is continued in Chapter 13, where the natural satellites and rings of these objects are treated in detail, with abundant use made of the missions to the outer planets. In Chapter 14, we discuss comets, beginning with an historical introduction that highlights the importance of comet studies to the development of modern astronomy. It summarizes the ground- and space-based imagery and discoveries, but makes use of earlier derivations to discuss cometary orbits. This chapter ends with the demise of comets and the physics of meteors. Chapter 15 treats the study of meteorites and the remaining small bodies of the solar system, the asteroids (*aka* minor planets, planetoids), and the outer solar system "Kuiper Belt" objects, and the closely related objects known as centaurs, plutinos, cubewanos, and others, all of which are numbered as asteroids. The chapter ends with discussions of the origin of the solar system and of debris disks around other stars, which point to widespread evidence of the birth of other planetary systems. Finally, in Chapter 16, we discuss the methods and results of extra-solar planet searches, the distinctions among stars, brown dwarfs, and planets, and we explore the origins of planetary systems in this wider context.

At the end of nearly every chapter we have a series of challenges. Instructors may use these as homework assignments, each due two weeks after the material from that chapter were discussed in class; *we* did! The general reader may find them helpful as focusing aids.

November 1, 2007 *E. F. Milone & W. J. F. Wilson*

Acknowledgments

These volumes owe their origin to more than 30 years of solar system classes in the Astrophysics Program at the University of Calgary, called, at various times, Geophysics 375, Astrophysics 301, 309, and 409. Therefore, we acknowledge, first, the students who took these courses and provided feedback. It is also a pleasure to thank the following people for their contributions:

David Mouritsen, formerly of Calgary and now Toronto, provided for Chapter 1 and our covers an image of his original work of art, an interpretation of Kepler's *Mysterium Cosmographicum*, in which the orbits of the planets are inscribed within solid geometric figures.

In Chapter 3, the Bradstreet and Steelman software package, *Binary Maker 3* was used to create an image to illustrate restricted three-body solutions.

University of Calgary Professor Emeritus Alan Clark gave us an image of an active region and detailed comments on the solar physics material of Chapter 4; Dr Rouppe van der Voort of the University of Oslo provided high-quality images of two other active region figures, for Chapter 4; the late Dr Richard Tousey of the US Naval Research Laboratory provided slides of some of the images, subsequently scanned for Chapter 4; limb-darkened spectral distribution plots were provided by Dr Robert L. Kurucz, of the Harvard-Smithsonian Center for Astrophysics; Dr. Charles Wolff, of Goddard Space Flight Center, NASA, reviewed the solar oscillations sections and provided helpful suggestions.

Dr. D. J. Stevenson provided helpful criticism of our lunar origins figures, and Dr. Robin Canup kindly prepared panels of her lunar simulations for our Fig. 8.10.

Dr Andrew Yau provided excellent notes as a guest lecturer in Asph 409 on the Martian atmosphere and its evolution, which contributed to our knowledge of the material presented in Chapters 9, 10, and 11; similarly, lectures by Professor J. S. Murphree of the University of Calgary illuminated the magnetospheric material described in Chapter 11.

NASA's online photo gallery provided many of the images in Chapters 8, 9, 12, 13, 14, and some of those in Chapter 15; additional images were provided by the Naval Research Laboratory (of both the Sun and the Moon). Some of these and other images involved work by other institutions, such as the U.S. Geological Survey, the Jet Propulsion Laboratory,

Arizona State Univ., Cornell, the European Space Agency, the Italian Space Agency (ASI), CICLOPS, CalTech, Univ. of Arizona, Space Science Institute, Boulder, the German Air and Space Center (DLR), Brown University, the Voyagers and the Cassini Imaging Teams, the Hubble Space Telescope, University of Maryland, the Minor Planet Center, Applied Physics Laboratory of the Johns Hopkins University, and the many individual sources, whether cited in captions or not, who contributed their talents to producing these images.

Dr. John Trauger provided a high resolution UV image of Saturn and its auroras for Chapter 12.

Dr William Reach, Caltech, provided an infrared mosaic image of Comet Schwassmann–Wachmann 3, and Mr John Mirtle of Calgary provided many of the comet images for Chapter 14, including those of Comets 109P/Swift–Tuttle, C/1995 O1 (Hale–Bopp), C/Hyakutake, Lee, C/Ikeya–Zhang, Brorsen–Metcalfe, and Machholz; Professor Michael F. A'Hearn of the University of Maryland for his critique of the comet content of Chapter 14.

Dr Allan Treiman, of the Lunar and Planetary Institute, Houston, was kind enough to provide background material on the debate over the ALH84001 organic life question, mentioned in Chapter 15. Mr Matthias Busch made available his *Easy-Sky* images of asteroid family distributions, and the Minor Planet Center provided the high-resolution figures of the distributions of the minor planets in the inner and outer solar system.

Dr Charles Lineweaver, University of New South Wales, provided a convincing illustration for the brown dwarf desert, illustrated in Chapter 16; University of Calgary graduate student Michael Williams provided several figures from his MSc thesis for Chapter 16.

Mr Alexander Jack assisted in updating and improving the readability of equations and text in some of the early chapters, and he and Ms Veronica Jack assisted in developing the tables of the extra-solar planets and their stars for Chapter 16.

In addition, we thank the many authors, journals, and publishers who have given us permission to use their figures and tabular material or adaptations thereof, freely. Finally, it is also a pleasure to thank Springer editors Dr Hans Koelsch, Dr Harry Blom, and their associate, Christopher Coughlin, for their support for this project.

Contents

Background Science and the Inner Solar System

1. Perceptions of the Solar System in History

The solar system has been around for a long time! Our perceptions of it, on the other hand, date back, arguably, only to the Upper Paleolithic (~70,000 to ~10,000 years ago).

The Paleolithic evidence for interest in the Moon, for example, is in the form of possible tallies of days in such features as the 17-in. high sculpture called the *Venus of Laussel* in a rock shelter dated from 20,000 to 18,000 y BP in the Dordogne region of France (see Campbell 1988, pp. 65–66 for a discussion of its symbolic significance and Marshack 1972, p. 335 for arguments for its use as a tally) and the Blanchard bone (among other artifacts) with a complicated chain of crescent incisions, also investigated by Marshack. The amply endowed *Venus* holds in her right hand an upturned horn on which are incised 13 grooves (defining 14 non-groove areas), a number said to represent an approximation to the number of waxing crescents in a year ($12\frac{1}{3}$) and the number of days between new and full moon ($\sim 14\frac{3}{4}$) (Marshack 1972).

In the Neolithic (or "New Stone Age," roughly from 6500 to 1500 BC), the evidence for the importance of the Sun and the Moon, at least, is overwhelming. It can be found in the many ancient alignment sites in the British Isles and is echoed in Stone Age cultures around the world, at least according to some interpretations. See Thom (1972) for a flavor of that evidence or, for example, Kelley and Milone (2005, Chapter 6) for a more recent summary.

Aside from practical astronomy, with calendrical usefulness for both agriculture and religions, astrology plays an increasing role in late antiquity (first several centuries AD). For the past four millennia, and especially during Hellenic and Hellenistic times (prior to and after Alexander the Great, respectively), we do know what people thought regarding the nature and origin of the solar system. Without attempting detailed examination of each one, we can characterize the principal theories about the solar system (in earlier times, the entire cosmos) in as shown in Table 1.1.

Most of the many notable figures in Tables 1.1 and 1.2 are discussed in Kelley and Milone (2005). Here, we single out only three for further discussion.

Martianus Capella was a poet and summarizer, who probably wrote his allegorical poem *Satyricon* after the sack of Rome by the Huns in 410 AD, which he mentions, and before 429 AD, when Carthage was overrun by Vandals, which he does not. His description of a quasi-heliocentric system

Table 1.1. Ancient theories of the solar system

Pythagoras (~6th BC): Earth at centre of planetary spheres (included Sun, Moon and the sphere of the fixed stars)
Anaxagoras of Clazomenae (c. 500–428 BC*)*: The Sun is a hot iron mass (based on meteorite evidence), bigger than the Peloponnesus; the Moon is a stone; the Earth, the centre of a cosmic vortex
Philolaus (~5th c. BC*):* The Earth moves, while the stellar sphere is immobile. He felt that there should be ten planets (fixed star sphere included), so Philolaus invented an anti-Earth, perpetually located between the Earth and a central fire about which all the planets, including the sun, moved; orbits were circular, but not coplanar. This is the earliest recorded theory to consider the earth as a moving object—but it was to explain the daily western movement of the "fixed" stars and other objects, not the annual motions of the Sun or the planets
Eudoxus of Cnidus (~408–355 BC*):* The fixed stars and each planet are carried on separate, concentric, rotating spheres, on various axes, centered on the Earth
Aristarchus of Samos (~250 BC*)*: The Sun is at the center; the Earth both revolves around the Sun and rotates on its own axis; the Moon revolves around the Earth
Apollonius of Perga (~220 BC*)*: Combinations of motion in circular orbits
Hipparchus (2nd c. BC*)*: The Earth is at the center; planets (including the Sun) revolve around the Earth; orbits are circular but non-concentric
Claudius Ptolemy (2nd c. AD*)*: The Earth is at the center; planets (including the Sun) revolve around the Earth; the orbits are combinations of circular motions, characterized by deferent orbits and epicyclic gyrations
Origen (3rd c. AD*)*: There exist a multiplicity of worlds, with the creation, fall, and redemption occurring on each
Martianus Capella ($\lesssim 5^{\mathrm{th}}$ *c. ?)*: His *Satyricon* refers to a Sun-centered solar system; this popular work kept the notion alive in the West to Copernicus' time
Aryabhata (b. 476 AD*)*: He allowed the possibility of a heliocentric universe (but his work was not known in the West until after Copernicus)

(like Tycho's model, it had Mercury and Venus orbiting the Sun), kept alive this idea. Copernicus explicitly mentions Capella's (and not Aristarchus') discussion of the heliocentric system.

Much later, following the Renaissance and Reformation, Tycho Brahe and Johannes Kepler were important transition figures.

Brahe himself contributed to a break in the classical paradigm by demonstrating, with observational data, that comets moved among the orbits of planets, thus shattering once and for all the notion that rotating crystalline spheres bore the planets. His discovery of a supernova and his determination that it was a very distant object demolished the idea of the immutability of the heavens. Moreover, this and his cometary discoveries refuted the ideas of Aristotle, for centuries, the highest authority on scientific questions.

Table 1.2. Post-medieval, pre-nineteenth century theories of the solar system

Nicholas of Cusa (1401–1464): He is said to have championed a Sun-centered theory; no explicit writings

Nicholas Copernicus (1473–1543): Sun-centered solar system; planets moved (as classically) in circular orbits

Tycho Brahe (1546–1601): Sun-centred planetary scheme—but Sun and planets revolve about the Earth

Johannes Kepler (1571–1630): Elliptical orbits; this is the first explicit departure from circular orbits. Keplerian empirical "laws"

René Descartes (1596–1650): The solar system is a complex of vortices; moons and planets arise from vortices within vortices

Isaac Newton (1642–1727): Planetary orbital motion due to gravity. The solar system is far from the stars (considered distant because of lack of parallax and relative motions) which are themselves, therefore, suns

Georges-Louis Leclerc Buffon (1707–1788): Collisional origin for the solar system (Sun with comet)

Immanuel Kant (1724–1794) and Simon de Laplace (1749–1827): The solar system had a nebular origin; contraction and conservation of angular momentum caused disk formation

Ernst Florenz Friedrich Chladni (1756–1827): The early aggregation of dust became planetesimals, and some of these, planets (Chladni 1794)

Kepler's early notion of the heliocentric planetary orbits carried on (crystalline) spheres inscribing and inscribed by the five regular polyhedral solids (see Figure 1.1) as expressed in the first half of his *Mysterium Cosmographicum* (Kepler 1596), evolved over his lifetime into a realization that the orbits were ellipses produced by forces that depended on the distance from the Sun. His persistence in trying to make sense of Tycho Brahe's highly precise data led to his conclusion that planetary orbits could not be circular. The consequences of this profound discovery resulted in the "Breaking of the Circle," in many ways (Nicholson 1950).

Thus their pursuit of the highest quality observational data and unflinching belief in the meaningfulness of those data led both of them to renounce the geocentric universe, although Brahe's was a last effort to incorporate the idea of a stationary Earth into a defensible model.

There are many nineteenth and twentieth century theories. Most of these theories involve either collisions or accretions or both. Table 1.3 presents some examples. Several of the theories, including the most recent, are cited in the references list. Note the trend from collisional theories to accretion theories in this interval.

Any thorough study of the solar system draws from chemistry, geology, and even biology, and numerous insights from those sciences will be brought into and used in this book. But, it is still basically astronomy. Observational

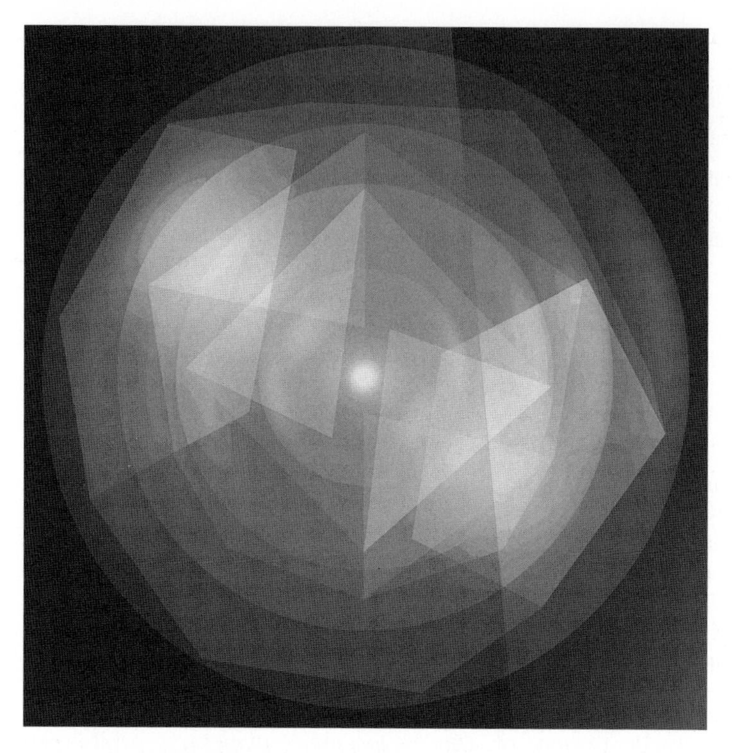

Fig. 1.1. An early Keplerian view of the solar system, inspired by Kepler's *Mysterium Cosmographicum* model of nested spheres and geometric solids. Original art by David Mouritsen (2005) and reproduced here with permission

astronomy has provided the basic data which are needed to understand the planets and other objects of the solar system, even if most of the new critical data now come from satellites and space probes, containing not only imaging cameras, but spectrographs (exploring the spectral energy distribution from radio and infrared to x-rays), magnetometers (to probe the structure and strength of magnetic fields), and particle detectors. But the latter does emphasize that remote sensing plays a vital role, and therefore our understanding of the solar system is more and more through "space science."

When we consider that the ultimate quest is to understand how the solar system came into being, how it evolves, and to what end, it is clear that a critical field of investigation has to be solar system dynamics. Therefore, this is an important area of study even for those who are not going to work for the Canadian, US, and European space agencies (CSA, NASA, and ESA, respectively) or for any of the space agencies developed in other countries around the world. We will take this subject up in a later chapter.

Table 1.3. Nineteenth and twentieth century theories of the origin of the solar system

A. W. Bickerton (1842–1929): Star–Sun collision; explosive eruption forms planets
R. A. Proctor (1837–1888): Planetesimal aggregations
T. C. Chamberlain (1843–1928): Star–Sun collision; tidal eruption creates planets; (1904): Planetesimals
F. R. Moulton (1872–1952): Star–Sun collision; tidal eruption + planetary accretion
K. O. B. Birkeland (1867–1912): Ions in solar atmosphere form rings in solar magnetic field
S. A. Arrhenius (1859–1927): Direct Sun–star collision, leaving the Sun and long filament as remnants
H. Jeffreys (1891–1989): Grazing Sun–star collision, leaving long filament that fragmented
J. H. Jeans (1877–1946): Star–Sun collision producing tidal filament
H. P. Berlage (1856–1934): Solar particle emission lead to gaseous rings/disks
H. N. Russell (1877–1957): Binary star component disrupted, forming a filament
D. ter Haar (1938): Contracting, turbulent solar envelope developed into planets
H. O. G. Alfven (1908–1995): Sun collided with a gas cloud which became ionized, and formed rings in the Sun's magnetic field; electromagnetic braking and transfer of angular momentum
O. I. Schmidt (1891–1956): Sun collided with a swarm of interstellar bodies which became planets by accretion; refined by R. A. Lyttleton (1911–1995)
C. F. von Weizsäcker (1912–2007): Turbulent eddies in protosun formed planets and satellites
F. Hoyle (1915–2001): Sun's binary companion went supernova, producing gaseous shells; remnant star left the system
F. Whipple (1906–2004): Protosun captured dust cloud of large angular momentum
G. Kuiper (1905–1973): Gravitational instabilities in protosun's gaseous envelope became planets
V. S. Safronov (1917–1999): Aggregation of dust into planetesimals
A. G. W. Cameron (1924–2005): Gaseous protoplanet theory
C. Hayashi et al. (b. 1920): Aggregation into planetesimals ("Kyoto" school)

The investigation of the nature of the solar system points to several striking facts:

- The solar rotation and the revolution of all the planets are in the same sense: CCW as viewed from the north ecliptic pole (NEP).
- The orbits are very nearly coplanar (the biggest departures being for the innermost and the outermost (usually) planets—Mercury and Pluto); and,

Table 1.4. The Titius–Bode law

$r = 2^n \cdot 0.3 + 0.4$

Planet	n	Prediction	True a
Mercury	$-\infty$	0.4	0.39
Venus	0	0.7	0.72
Earth	1	1.0	1.00
Mars	2	1.6	1.52
Minor planets	3	2.8	$< 2.8 >$
Jupiter	4	5.2	5.20
Saturn	5	10.0	9.54
Uranus	6	19.6	19.18
Neptune	7	38.8	30.07
Pluto	8	77.2	39.46
Eris	9	154.0	67.78

again except for Mercury and dwarf and minor planets, very nearly circular; the spacing of the planets is not random, but is described by the Titius–Bode law.[1]

- Although the mass is strongly concentrated in the Sun, the angular momentum is not.
- The coplanar revolutions of the planets and the solar rotation (to be discussed later) already make a disk formation of the solar system more likely than a collisional origin.
- The low orbital eccentricities of the planets strengthen the case. The circularity of Neptune's orbit, the outermost and thus least strongly bound of all the major planets (Pluto, Eris, and other "dwarf planets" excepted from this category), is especially compelling.

These and other properties of the solar system will be reviewed at the beginning of Chapter 3 and again later, mainly in chapters 15 and 16 of Milone & Wilson (2008), when we consider the solar system's origins.

[1] See Table 1.4. The Titius–Bode law (Bode 1772; Wurm 1787; Jaki 1972; Nieto 1972) can be expressed in the form:

$$r = (3 \times 2^n + 4)/10 \tag{1.1}$$

where $n = -\infty$, 0, 1, 2, 3, ... 6 (7–9, Neptune–Eris, are not well represented). The relation can be better expressed in the more modern form,

$$r_n = r_0 \, a^n \tag{1.2}$$

where r_n is the distance in AUs of the nth planet (Mercury is $n = 0$), in order of distance, from the Sun, and where $a \equiv 1.73$ (in the Blagg–Richardson formulation; see Nieto 1972). Note that one can either determine or assume the quantity r_0.

In our final chapter, we take up the properties of extrasolar planets. The existence of disks around other stars and giant molecular clouds with which protostars are associated are further evidence for a disk origin of the solar system. Currently, planets are believed to arise from protostellar disks, but there are sharp disagreements over the separate roles of disk condensation and accretion of other condensates or larger—perhaps pre-existing clumps of matter. The importance of disks is empirically based not only on the observations of infrared tori seen around other stars (most famously, but far from exclusively, β Pictoris), but also on the basis of meteorite and theoretical studies. However, theoretical difficulties in explaining the formation of the lesser giants at their current locations in the solar system and the presence of "hot Jupiters" in other star systems strongly suggest dynamical migrations of planets from their points of origin, if the disk origin is to be sustained. Indeed, modern simulations provide mounting evidence that the lesser giants in the solar system, Uranus and Neptune, were formed closer to the Sun and were driven further out by dynamical interactions. In the last chapter of Milone & Wilson (2008) we will summarize what can be generalized about the origins of planetary systems and the prospects of finding terrestrial planets in other star systems.

Any study of the origin of the solar system must be a kind of mystery-solving expedition. So, we need to take note of the clues as we go along, as a police inspector in attempting to unscramble a forensic puzzle.

We start by providing some basic investigatory tools and then begin the search for clues in the dynamical and physical structure of the solar system.

References

Bode, J. E. 1772. (*Deutliche*) *Anleitung zur Kenntniss des gestirnten Himmels*, 2nd Ed. (Hamburg: Dieterich Anton Harmsen), pp. 462–463. (Cited in Jaki 1972).

Cameron, A. G. W. 1978. in *The Origin of the Solar System*, ed. S. F. Dermott (New York: Wiley), p. 49.

Campbell, J. 1988. *Historical Atlas of World Mythology Vol. I: The Way of the Animal Powers* (New York: Harper & Row).

Chamberlain, T. C. 1904. *Carnegie Inst. Washington Yearbook*, **3**, 133.

Chladni, E. F. F. 1794. *Über den Ursprung der von Pallas Gefundenen und anderer ihr ähnlicher Eisenmassen*, ed. J. F. Hartknoch (Riga: Repr. by Meteoritic Society, 1974), p. 56.

Hayashi, C., Nakazawa, K., and Nakagawa, Y. 1985 in *Protostars and Planets II*, Ed. D. C. Black and M. S. Matthews (Tucson: University of Arizona Press), p. 1100.

Jaki, S. L. 1972. "The Early History of the Titius–Bode Law," *American Journal of Physics*, **40**, 1014–1023.

Kelley, D. H. and Milone, E. F. 2005. *Exploring Ancient Skies: An Encyclopedic Survey of Archaeoastronomy* (New York: Springer Verlag).
Kepler, J. 1596. *Mysterium Cosmographicum*, 1981 translation by A. M. Duncan, Ed. E. J. Aiton (New York: Abaris Books).
Kuiper, G. P. 1951. "On the Origin of the Solar System" in *Astrophysics*, Ed. J. A. Hynek (New York: McGraw Hill), pp. 365–424.
Marshack, A. 1972. *The Roots of Civilization* (New York: McGraw Hill).
Milone, E. F., and Wilson, W. J. F. 2008. *Solar System Astrophysics: Planetary Atmospheres and the Outer Solar System* (New York: Springer).
Moulton, F. R. 1905. "Evolution of the Solar System," *Astrophysical Journal*, **22**, 166.
Nicholson, M. H. 1950. *The Breaking of the Circle: Studies in the Effect of the "New Science" upon Seventeenth Century Poetry* (Evanston: Northwestern University Press).
Nieto, M. M. 1972. *The Titius–Bode Law of Planetary Distances* (Oxford: Oxford University Press).
Proctor, R. A. 1898. *Other Worlds than Ours* (New York: D. Appleton), p. 220.
Russell, H. N. 1935. *The Solar System and Its Origin* (New York: MacMillan), pp. 95–96.
Safronov, V. S. 1969. *Evolution of the Protoplanetary Cloud and Formation* (Moscow: Nauka) (Tr. for NASA and NSF by Israel Prog. for Scientific Translations).
Thom, A. 1972. *The Megalithic Lunar Observatories of Britain* (Oxford: Oxford University Press).
von Weizsäcker, C. F. 1943. "Über die Entstehung des Planetensystems," *Zeitschrift für Astrophysik*, **22**, 319.
Wurm, J. F. 1787. *Astronomisches Jahrbuch fur das Jahr 1790* (Berlin: G. J. Decker), pp. 161–171 (Cited in Jaki 1972).

Challenges

[1.1] Try to categorize the theories of the origin of the solar system, listing the theories below each category. Is there any evidence for historical evolution or evidence of progress among the theories of a particular type?

[1.2] Compare the computed distances of the Titius–Bode and Blagg–Richardson laws to the mean distances of the planets from the Sun. A spreadsheet is the most convenient way of doing this. Can you formulate another relation that describes these distances precisely? (Hint: think non-linear. You can make use of a software package such as ***Tablecurve***[2] to find other relationships.)

[2] Tablecurve 2D Automated Curve Fitting Software v2.0 1994 ed., copyrighted by AISN Software Incorporated.

2. Basic Tools and Concepts

In this chapter, from the Greeks (through much subsequent development), we derive the tools of spherical astronomy. We will describe the basic theorems of spherical trigonometry and emphasize the usefulness of the sine and cosine laws. We will also describe the ellipse and its properties, in preparation for a subsequent discussion of orbits.

2.1 Circular Arcs and Spherical Astronomy

All astronomical objects outside the solar system are sufficiently far away that their shifts in position due to parallax (caused by periodic motions of the Earth) and proper motion (caused mainly though not exclusively by the objects' own motion) are too small to be discerned—at least by the unaided eye. Historically, this suggested that these objects could be regarded as being fixed to a sphere, the *celestial sphere*, of some very large radius centered on the Earth. Objects within the solar system change position with time, e.g., a superior planet's orbital motion causes an eastward motion across our sky relative to the distant stars and the Earth's motion causes the planet to follow a retrograde loop. However, objects within the solar system can be referenced to the celestial sphere at any given instant of time.

We may wish to calculate the distance measured across the sky from one object to another knowing the distance of each of them from a third object, and also knowing an appropriate angle. ("Distance across the sky" is actually an arclength, measured in units of angle such as degrees or radians.) In doing this, we are in essence drawing arcs joining three objects to form a triangle on a spherical surface, so the mathematical relationships involved are those of spherical trigonometry. When we apply them to the sky we are practicing *spherical astronomy.* We note also that the objects do not need to be real; one or more of them can be a reference point, such as the north or south celestial pole.

A *spherical triangle* is a triangle on the surface of a sphere such that each side of the triangle is part of a *great circle*, which has as its center the center of the sphere (a *small circle* will have its center along a radius of the sphere). For example, the Earth's equator is a great circle (assuming a spherical Earth), and any line of latitude other than the equator is a small circle.

The procedure in spherical astronomy lies primarily in the calculation of one side or angle in a spherical triangle where three other appropriate quantities are known, e.g., we may wish to find the length (in degrees) of one side of a spherical triangle given two other sides and the angle formed between them or find one side given a second side and the angle opposite each side or find one angle given a second angle and the side opposite each angle.

First, we review the basics of spherical trigonometry and then derive the cosine and sine laws, analogous but not identical to the cosine and sine laws of plane geometry. Additional theorems relating the three angles and three sides (involving, for example, haversines[1]) can also be found, but will not be derived here; for such theorems, see Smart's (1977) or Green's (1985) spherical astronomy texts, for example.

Figure 2.1(a) shows an example of a spherical triangle. The three ellipses are great circles seen in projection, and the spherical triangle (marked by heavy lines) is formed by their intersections. In Figure 2.1(b), we label the three sides of the triangle a, b, and c, and the angles at the three corners A, B, and C. We also draw a radius from the center of the sphere to each corner of the triangle (dashed lines).

Figure 2.2(a) shows an enlarged view of the spherical triangle and the radii to the center of the sphere. A circle is coplanar with its radii, so the shaded cross-section in this figure is a plane triangle (i.e., with straight sides).

It is important to distinguish between angles A, B, and C, which are the angles seen by an (two-dimensional) observer on the surface of the sphere and angles a, b, and c, which are the angles seen at the center of the sphere (Figure 2.2(b)). That is, angle A is the angle (measured in degrees or radians)

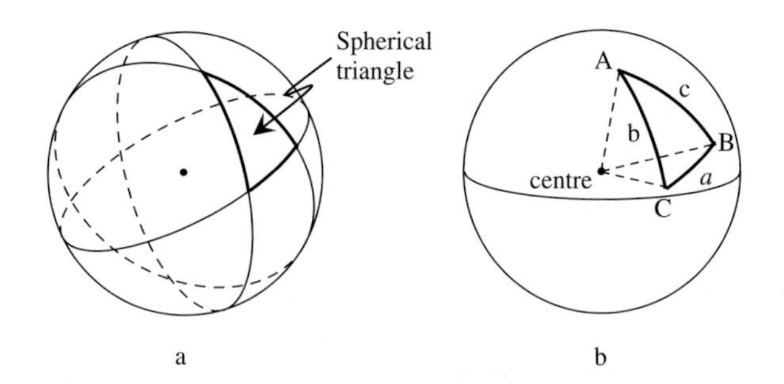

Fig. 2.1. The sphere with a spherical triangle on its surface

[1] hav $\theta = 1/2\,[1 - \cos\theta]$.

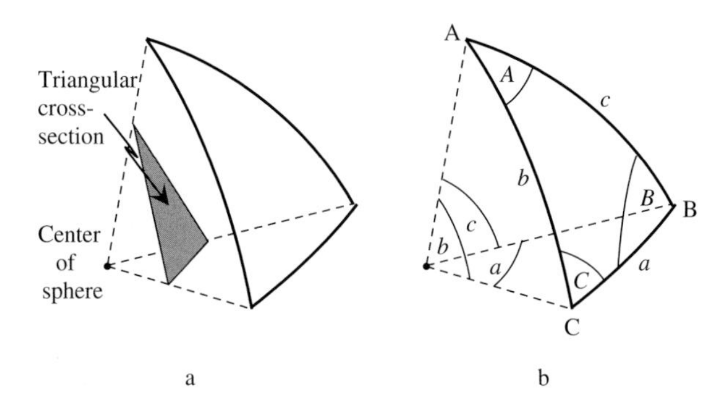

Fig. 2.2. Relating the sides of a spherical triangle to angles a, b, and c at the center of the sphere and angles A, B, and C on the surface of the sphere. The *heavy lines* are arcs of great circles; the *dashed lines* are radii of the sphere

at the intersection of sides b and c on the surface of the sphere, whereas angle a is the angular separation "across the sky" from point B to point C (also measured in degrees or radians) as viewed from the center of the sphere. The principal equations relating these quantities to each other are the cosine and sine laws of spherical astronomy, which we will now derive.

NB: Small circle arcs have a different relation to interior angles. The laws derived here, therefore, do not apply to them.

2.1.1 The Law of Cosines for a Spherical Triangle

In this section we demonstrate a proof of the cosine law:

$$\cos a = \cos b \cos c + \sin b \ \sin c \ \cos A$$

First, we derive some useful results for a right spherical triangle ($\angle C = 90°$ in Figure 2.3), which will also be useful in proving the sine law.

Take the radius of the sphere to be one unit of distance (OA = 1).

Note that any three points define a flat plane, so planes OAC, OAB, and OBC are each flat, and plane OAC is perpendicular to plane OBC. (To visualize this, it may help to think of arc BC as lying along the equator, in which case arc AC is part of a circle of longitude and arc AB is a "diagonal" great circle arc joining the two. Circles of longitude meet the equator at right angles, and longitudinal planes are perpendicular to the equatorial plane.)

From A, drop a line AD perpendicular to the plane OBC. Plane OAC $\perp$ plane OBC, so point D is on the line OC. From D, draw a line DE $\perp$ OB. (DE is, of course, not $\perp$ OC.) Then:

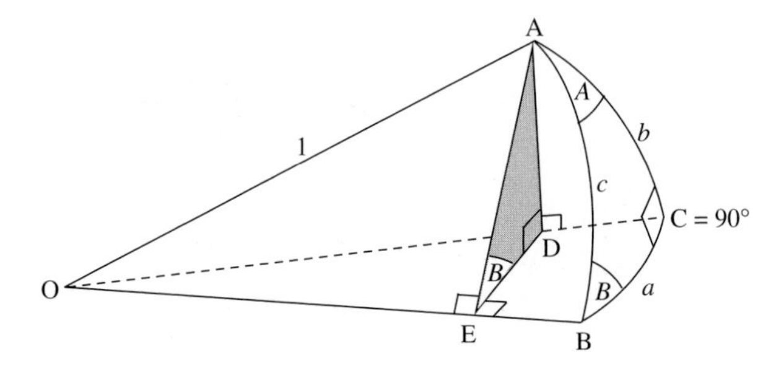

Fig. 2.3. Projection of a right spherical triangle onto a plane perpendicular to the base plane OBC. For convenience, we define points A and B to be located at the apices of angles A and B, respectively

Plane ADE $\perp$ plane OBC because line AD $\perp$ plane OBC
Plane ADE $\perp$ line OB because DE $\perp$ OB and AD $\perp$ plane OBC
$\angle$ADE $= 90°$ because plane OAC $\perp$ plane OBC
$\angle$OED $= 90°$ and $\angle$OEA $= 90°$ because plane ADE $\perp$ line OB

Now define a plane tangent to the sphere (i.e., $\perp$ to line OB) at point B; then $\angle B$ lies in this tangent plane and, in fact, is equal to the angle of intersection of the two planes OAB and OBC. Plane ADE is parallel to the tangent plane, since both planes are $\perp$ line OB, so it follows that:

$$\angle \mathrm{AED} = \angle B$$

Then

$$\sin c = \frac{\mathrm{AE}}{\mathrm{OA}} = \frac{\mathrm{AE}}{1} = \mathrm{AE}$$
$$\cos c = \frac{\mathrm{OE}}{\mathrm{OA}} = \mathrm{OE}$$
$$\sin b = \frac{\mathrm{AD}}{\mathrm{OA}} = \mathrm{AD}$$
$$\cos b = \frac{\mathrm{OD}}{\mathrm{OA}} = \mathrm{OD}$$

These sides are shown in Figure 2.4. Then from ΔODE,

$$\cos a = \frac{\mathrm{OE}}{\mathrm{OD}} = \frac{\cos c}{\cos b}$$

or

$$\cos c = \cos a \;\cos b \tag{2.1}$$

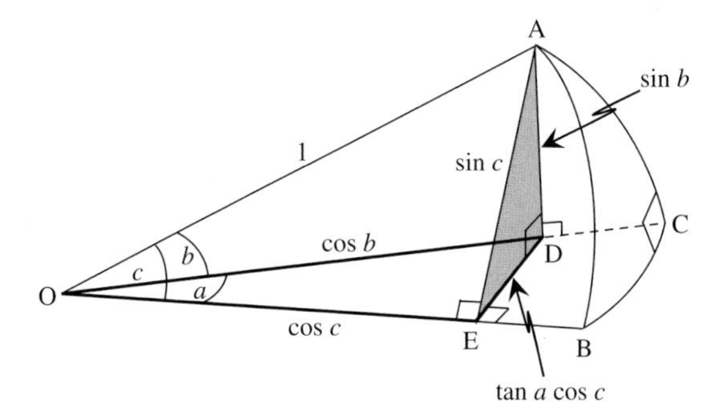

Fig. 2.4. Lengths of relevant sides, taking the radius of the sphere to be one unit of distance (OA = 1)

Also from ΔODE,

$$\tan a = \frac{\mathrm{DE}}{\mathrm{OE}} = \frac{\mathrm{DE}}{\cos c}$$

or

$$\mathrm{DE} = \tan a \ \cos c$$

Therefore, from ΔADE,

$$\cos B = \frac{\mathrm{DE}}{\mathrm{EA}} = \frac{\tan a \ \cos c}{\sin c} = \tan a \ \cot c$$

Now $\angle C$ has been defined to be a right angle, but there is nothing to distinguish $\angle A$ from $\angle B$. It follows that any rule derived for the left-hand triangle in Figure 2.5, below, has to be equally true for the right-hand triangle.

Thus, any rule derived for $\angle B$ is equally true for $\angle A$ with suitable relettering:

$$\cos A = \tan b \ \cot c \tag{2.2}$$

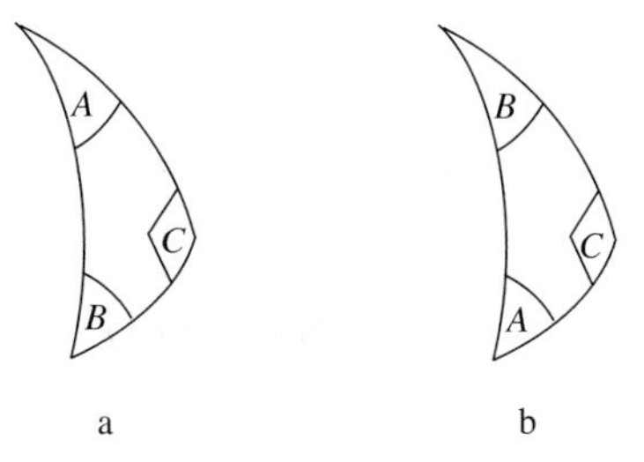

Fig. 2.5. Un-handedness of spherical triangles

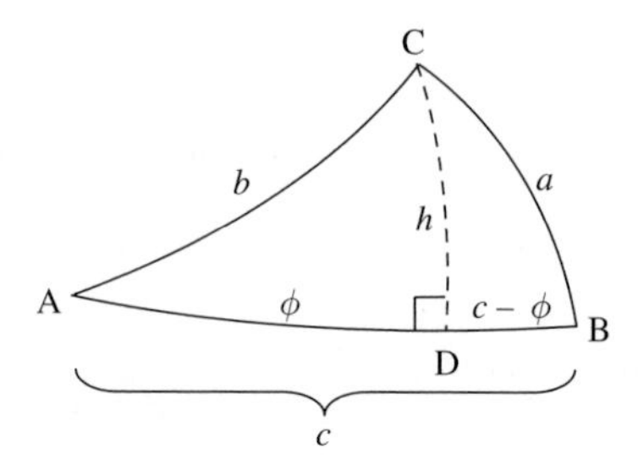

Fig. 2.6. The general spherical triangle as a combination of two right spherical triangles

(One could derive this directly by redrawing Figure 2.3 with the plane passing through point B perpendicular to the line OA instead of through point A perpendicular to the line OB.) Another equation can be obtained easily from ΔADE in Figure 2.4:

$$\sin B = \frac{\text{AD}}{\text{AE}} = \frac{\sin b}{\sin c}$$

$$\therefore \ \sin b = \sin c \ \sin B \tag{2.3}$$

With equations (2.1) to (2.3) in mind, we can now look at the general spherical triangle (no right angles), as shown in Figure 2.6.

Drop an arc $h \perp$ arc AB from point C to point D in Figure 2.6. This divides the triangle into two right spherical triangles.

Define arc AD to take up an angle ϕ as seen from the center of the sphere (point O in Figure 2.4). Then arc DB takes up angle $(c - \phi)$.

Apply equation (2.1) to each right spherical triangle in Figure 2.6; then,

ΔADC: side b is opposite to the right angle, so,

$$\cos b = \cos h \ \cos \phi \tag{2.4}$$

ΔBDC: side a is opposite to the right angle, so,

$$\cos a = \cos h \ \cos (c - \phi) \tag{2.5}$$

Divide equation (2.5) by equation (2.4) and use the standard trigonometric identity, $\cos(c - \phi) = \cos c \ \cos \phi + \sin c \ \sin \phi$, to get

$$\frac{\cos a}{\cos b} = \frac{\cos h \ \cos (c - \phi)}{\cos h \ \cos \phi} = \frac{\cos (c - \phi)}{\cos \phi}$$

$$= \frac{\cos c \ \cos \phi + \sin c \ \sin \phi}{\cos \phi} = \cos c + \sin c \ \tan \phi$$

or

$$\cos a = \cos b \ \cos c + \cos b \ \sin c \ \tan \phi \tag{2.6}$$

Now use equation (2.2) to obtain an equation for tan ϕ:

$$\cos A = \tan \phi \ \cot b = \frac{\tan \phi \ \cos b}{\sin b}$$

or

$$\tan \phi = \frac{\cos A \ \sin b}{\cos b} \tag{2.7}$$

Substituting (2.7) into (2.6), we arrive at,

$$\cos a = \cos b \ \cos c + \sin b \ \sin c \ \cos A. \tag{2.8}$$

Equation (2.8) is the cosine law for spherical triangles. Because none of the angles in the triangle are right angles, there is nothing to distinguish one angle from another, and the same equation has to apply equally to all three angles:

$$\begin{aligned}
\cos a &= \cos b \ \cos c + \sin b \ \sin c \ \cos A \\
\cos b &= \cos c \ \cos a + \sin c \ \sin a \ \cos B \\
\cos c &= \cos a \ \cos b + \sin a \ \sin b \ \cos C
\end{aligned}$$

Note the canonical rotation of angles from one formula to the next.

2.1.2 Law of Sines for a Spherical Triangle

Application of equation (2.3) to ΔADC and ΔBDC in Figure 2.6 gives, respectively,

$$\begin{aligned}
\sin h &= \sin b \ \sin A \quad \text{and} \\
\sin h &= \sin a \ \sin B
\end{aligned}$$

where sides b and a in Figure 2.6 are opposite the right angles, and so replace side c in equation (2.3). Therefore,

$$\begin{aligned}
&\sin b \ \sin A = \sin a \ \sin B \quad \text{and} \\
&\frac{\sin a}{\sin A} = \frac{\sin b}{\sin B}
\end{aligned}$$

Again there is nothing to distinguish one angle from another, so this equation also has to apply equally to all angles:

$$\frac{\sin\ a}{\sin\ A} = \frac{\sin\ b}{\sin\ B} = \frac{\sin\ c}{\sin\ C} \tag{2.9}$$

Equation (2.9) is the law of sines for spherical triangles.

2.1.3 Other Laws

Two other formulae which may be useful in particular cases are the *analogue formula* of Smart (1977, p. 10):

$$\sin\ a\ \cos\ B = \cos\ b\ \sin\ c - \sin\ b\ \cos\ c\ \cos\ A \tag{2.10}$$

and Smart's (1977, p. 12) *four-parts formula*,

$$\cos\ a\ \cos\ C = \sin\ a\ \cot\ b - \sin\ C\ \cot\ B \tag{2.11}$$

The quantities in each of these may be canonically rotated, as per the cosine and sine laws.

2.1.4 Applications

Uses for spherical trigonometry abound. One example is to use a terrestrial system triangle to find the length of a great circle route for a ship or plane given the initial and final points of the route. The terrestrial coordinate system, (λ,ϕ), consists of latitude, ϕ, longitude, λ, the equator, the poles, and the sense in which latitude and longitude are measured (N or S from the equator and E or W from Greenwich, the "prime meridian," resp.). The units of longitude may be in hms or degrees; latitude is always given in degrees. The two longitude arcs from the north or south pole to the initial and final points then form two sides of a spherical triangle, and the great circle route forms the third side.

A spherical triangle also can be used to find the arc length between two objects in the sky with a coordinate system appropriate to the sky, or to transform between two coordinate systems. Several different coordinate systems are in use. In the *horizon* or *altazimuth (A,h)* system, the coordinates are *altitude, h,* measured along a vertical circle positive toward the zenith from the horizon, and *azimuth, A,* measured from a fixed point on the horizon, traditionally the North point, CW around toward the East. Both are measured in degrees of arc. For example, an observer can use a theodolite to observe the altitude and azimuth of a star, then use these to find the latitude of the observing site.

For astronomical applications, corrections need to be made for the effect of the Earth's atmosphere on the altitude: refraction by the atmosphere raises the altitude, h, by a value which depends itself on the altitude. At the horizon, the correction is large and typically amounts to $\sim 34'$. At altitudes above about $40°$, and expressed in terms of the zenith distance ($\zeta = 90° - h$), the difference in altitude is $\sim 57.3'' \tan \zeta$. A measured altitude must be decreased by this amount to obtain the value of the altitude in the absence of the atmosphere. Another correction must be made for the "dip" of the horizon when the observer is not at ground or sea level (for example, when the observer is on the bridge of a naval vessel).

The *equatorial system* of astronomical coordinates has two variants:

The (H or HA, δ) system, which uses *Hour Angle*, H, and *declination*, δ
The (α,δ) system, which uses *Right Ascension*, α, and declination

Thus, the location of astronomical objects such as the Sun or stars in the sky depends on the observer's latitude, the declination (distance above the celestial equator), δ, of the object, and the time of day.

Declination is measured North (+) or South (−) toward the N or S Celestial Poles from the *celestial equator*, the extension of the Earth's equator into the sky. Both H and α are measured in units of time. H is measured positive westward from the observer's meridian; α is measured eastward from the Vernal Equinox (the ascending node of the *ecliptic*; see below). Note that the (H,δ) system is dependent on the site, because H at any instant depends on the observer's longitude; the (α,δ) system is essentially independent of the observer's location. The latter is of use, for example, for a catalog of stars or other relatively 'fixed' objects.

At a particular site, H increases with time (the hour angle of the Sun + 12^h defines the apparent solar time) and this causes both altitude and azimuth to change with time also. Thus, the equatorial systems are more fundamental coordinate systems for celestial objects than the altazimuth system. The connection between the two variants of the equatorial system is the sidereal time, Θ:

$$\Theta = H + \alpha \tag{2.12}$$

The origin of the (α, δ) equatorial system is the Vernal Equinox, symbolized by the sign of Aries, ♈, so the right ascension of this point is 0. Therefore sidereal time may be defined as,

$$\Theta \equiv H(\Upsilon) \tag{2.13}$$

(the hour angle of the Vernal Equinox). Recall that the Sun does not move along the celestial equator, but along the ecliptic, causing the different durations of sunshine with season (and latitude). Other consequences of this motion are the *equation of time* and the amplitude (maximum variation of the azimuth of the rising/setting Sun from the East/West points, respectively). See Figure 2.7 for illustrations of the (A,h) and (H,δ) systems and the quantities needed to compute one set of coordinates from the other. In this figure, ϕ is the observer's latitude, which is equal to the altitude of the celestial pole above the observer's horizon and to the declination of the observer's zenith.

The *ecliptic coordinate system*, (λ,β), involves the coordinates *celestial latitude*, β, and *celestial longitude*, λ, analogous to both the terrestrial coordinates (λ,ϕ) and the equatorial system (α,δ). The closer analogy, despite the names, is to the latter because the celestial longitude is measured CCW (viewed from the N) and from the same zero point, the Vernal Equinox, ♈. The two reference circles, the celestial equator and the ecliptic, intersect at the vernal and autumnal equinoxes. The angle between them, known as the *obliquity of the ecliptic*, ε, is about 23.440° at present—it is slowly decreasing with time. The ecliptic system is very important for solar system studies and for celestial mechanics, both of which deal primarily with the solar

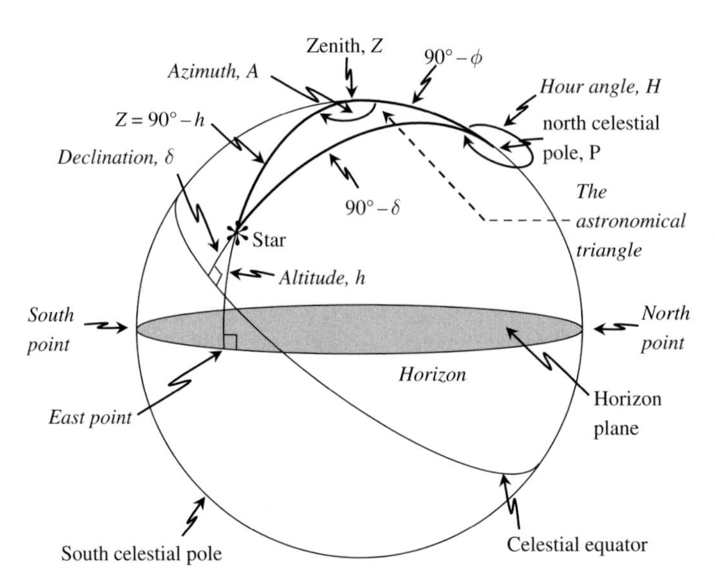

Fig. 2.7. The horizon A,h and equatorial H,δ coordinate systems. The spherical triangle whose apices lie at the star, the zenith, and the visible celestial pole is referred to as *the astronomical triangle*. Angle Z is the zenith angle of the star. Note that H is measured CW from south whereas A is measured CW from north

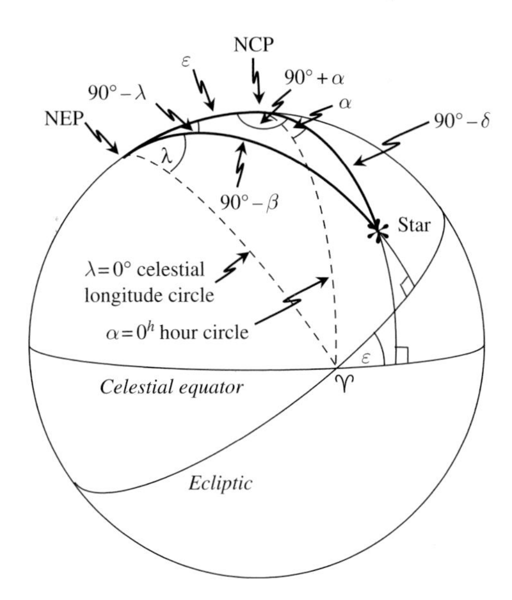

Fig. 2.8. The (λ,β) ecliptic and (α,δ) equatorial coordinate systems. ε is the obliquity of the ecliptic, i.e., the angle between the ecliptic and the celestial equator, and therefore also between the north ecliptic pole (NEP) and the north celestial pole (NCP). The spherical triangle involving the star, the NEP, and the NCP is used to obtain the transformation equations between the systems

system, the overlap with stellar kinematics, and dynamics notwithstanding. The relationship between the ecliptic and equatorial systems can be seen in Figure 2.8, which shows the angles needed to compute one set of coordinates given the other.

The transformation equations between systems are readily obtained by drawing both on a celestial sphere and solving the resulting spherical triangles for the unknown pair of coordinates. Thus to compute the ecliptic longitude and latitude, draw the equatorial (RA) and ecliptic systems on the celestial sphere and use the separation of the poles of the system as one of the triangle legs. For this purpose, the spherical sine and cosine laws are perfectly adequate, even for checking the quadrant of the longitudinal coordinate (which may be in any of the four quadrants).

Example 2.1

An astronomer wants to observe a particular star with a telescope on an altazimuth mount. A star atlas provides the star's α and δ; and the star's hour angle is then given by $H = \Theta - \alpha$ from (2.12), where Θ is the sidereal time. However, because the mount is altazimuth, the coordinates actually needed are the altitude and azimuth. Find the transformation equations to convert the star's coordinates from H and δ to A and h. (Note how H and A are defined, and be careful with signs in equations containing trigonometric functions.)

Solution to Example 2.1

Figure 2.7 shows the two relevant systems of coordinates. Here, ϕ is the observer's latitude and is equal to the declination of the observer's zenith. The star, the zenith, and the NCP form a spherical triangle referred to as the *astronomical triangle*, with sides $90° - h$ opposite the angle $360° - H$; $90° - \delta$ opposite the azimuth angle, A; and $90° - \phi$ opposite the angle formed at the star. Equations (2.8) and (2.9) then give, respectively,

$$\cos(90° - h) = \cos(90° - \phi)\cos(90° - \delta) + \sin(90° - \phi)\sin(90° - \delta)\cos(360° - H)$$

$$\sin A = \sin(90° - \delta)\sin(360° - H)/\sin(90° - h).$$

The identities,

$$\sin(90° - \theta) = \cos\theta$$
$$\cos(90° - \theta) = \sin\theta$$
$$\sin(360° - \theta) = -\sin\theta$$
$$\cos(360° - \theta) = \cos\theta$$

then give the transformation equations,

$$\sin h = \sin\phi\ \sin\delta + \cos\phi\ \cos\delta\ \cos H \tag{2.14}$$

$$\sin A = -\cos\delta\ \sin H/\cos h \tag{2.15}$$

Example 2.2

At some time of night, an observer at latitude 30°N sees a star at an altitude of 20° and an azimuth of 150°. Find the star's (H,δ) equatorial coordinates.

Solution

Now we need the equations for the inverse of the transformation in Example 2.1, i.e., from the horizon to the equatorial system. Using Figure 2.7 and the procedure in Example 2.1 again these are, from the cosine law,

$$\sin\delta = \sin\phi\ \sin h + \cos\phi\ \cos h\ \cos A \tag{2.16}$$

and, from the sine law,

$$\sin H = -\cos h\ \sin A/\cos\delta \tag{2.17}$$

Then with the values $\phi = 30°$, $h = 20°$, and $A = 150°$, equation (2.16) yields

$$\sin\delta = -0.53376, \text{ so that } \delta = -32.260°$$

and from (2.17), $\sin(H) = -0.55562$, so

$$H = -33.753° \text{ or } -33.753°/(15°/\text{h}) \cong -02^h\ 15^m = 02^h\ 15^m \text{ East.}$$

Example 2.3

What is the angular distance across the sky from Deneb (α Cygni) at $\alpha = 20^{\text{h}}\ 40^{\text{m}}\ 24^{\text{s}}$, $\delta = +45°\ 10'$ to Sirius (α Canis Majoris) at $\alpha = 6^{\text{h}}\ 43^{\text{m}}\ 48^{\text{s}}$, $\delta = -16°\ 41'$?

Solution

Figure 2.9 illustrates the (α,δ) equatorial coordinate system, the relevant spherical triangle, and the angles involved. Subscripts D and S signify Deneb and Sirius, respectively. We want to find the arc length of side a.

The arc lengths of sides b and c are

$$b = 90° - \delta_{\text{S}} = 90° - (-16°\ 41') = 106°\ 41' = 106.68°$$
$$c = 90° - \delta_{\text{D}} = 90° - 45°\ 10' = 44°\ 50' = 44.83°$$

There are several ways of expressing the angle A, all of which are equivalent by the fact that $\cos\theta = \cos(-\theta) = \cos(360° - \theta)$. Here we take the smallest

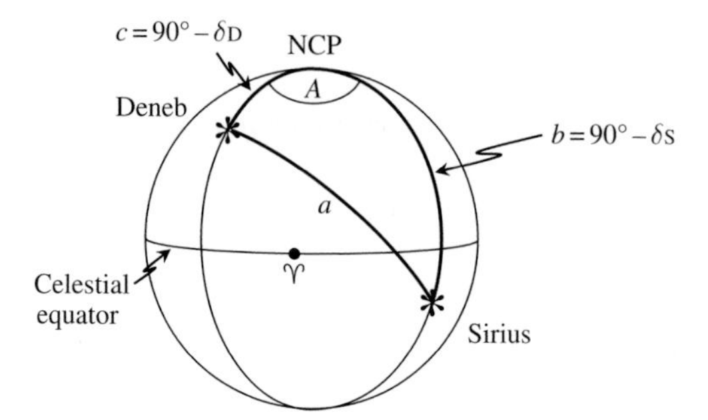

Fig. 2.9. The (α,δ) equatorial coordinate system, and the spherical triangle relating Deneb, Sirius, and the NCP. The RA arcs of the two stars are shown; angle A is the difference in right ascension between these two arcs

positive value of A, but it would be equally correct and perhaps simpler to take $A = \alpha_{\mathrm{S}} - \alpha_{\mathrm{D}}$ to obtain a negative angle or $A = \alpha_{\mathrm{D}} - \alpha_{\mathrm{S}}$ to obtain a positive angle $> 180°$.

From Figure 2.9,

$$\begin{aligned} A &= (24^{\mathrm{h}} - \alpha_{\mathrm{D}}) + \alpha_{\mathrm{S}} \\ &= (24^{\mathrm{h}} - 20^{\mathrm{h}}\ 40^{\mathrm{m}}\ 24^{\mathrm{s}}) + 6^{\mathrm{h}}\ 43^{\mathrm{m}}\ 48^{\mathrm{s}} \\ &= 10^{\mathrm{h}}\ 3^{\mathrm{m}}\ 24^{\mathrm{s}} \\ &= (10^{\mathrm{h}} \times 15°/\mathrm{h}) + (3^{\mathrm{m}} \times 1/60\ \mathrm{h/m} \times 15°/\mathrm{h}) \\ &\quad + (24^{\mathrm{s}} \times 1/3600\ \mathrm{h/s} \times 15°/\mathrm{h}) \\ &= 150.85°. \end{aligned}$$

Then, using the cosine law to find side a,

$$\begin{aligned} \cos a &= \cos b\ \cos c + \sin b\ \sin c\ \cos A \\ &= \cos\ 106.68°\ \cos\ 44.83° + \sin\ 106.68°\ \sin\ 44.83°\ \cos\ 150.85° \\ &= -0.7934. \end{aligned}$$

The inverse cosine is double-valued, so, within round-off error,

$$a = \cos^{-1}(-0.7934) = 142.6° \text{ or } 217.4°.$$

The shortest angular distance between any two objects on a sphere is always $\leq 180°$, so the distance across the sky from Deneb to Sirius is 142.6°.

See Schlosser et al. (1991/1994) or Kelley and Milone (2005) for more details on and worked examples of transformations.

2.2 Properties of Ellipses

Spherical astronomy is a very important tool in solar system astronomy, as well as other areas, but it is not, of course, the only one. We now turn from a consideration of the circular and spherical to review basic properties of ellipses. This is useful for understanding the orbits of solar system objects, considered in Chapter 3.

An ellipse is the locus of points $P(x, y)$, the sum of whose distances from two fixed points is constant.

That is, in Figure 2.10,

$$\ell_1 + \ell_2 = \text{constant}. \tag{2.18}$$

The two fixed points are the *foci* of the ellipse (singular: *focus*). Relevant geometric definitions of the quantities in Figures 2.10 and 2.11 are:

$a =$ semi-major axis (therefore the length of the major axis is $2a$);
$b =$ semi-minor axis (therefore the length of the minor axis is $2b$);
$f =$ distance of each focus from the center of the ellipse;
$r =$ distance from one focus to a point on the ellipse (e.g., point P);
$r_{\text{min}} =$ distance from either focus to the nearest point on the ellipse;
$r_{\text{max}} =$ distance from either focus to the farthest point on the ellipse.

We can now evaluate the constant in equation (2.18), above. We do this by noting that (2.18) is true for every point on the ellipse. It is therefore also true if we move point P to the right end of the ellipse (Figure 2.12) so that ℓ_1 and ℓ_2 lie along the major axis. The foci are symmetrically placed, so ℓ_2 equals the distance from the left focus to the left end of the major axis and ℓ_1 and ℓ_2 add up to $2a$ for this point.

Since $\ell_1 + \ell_2 =$ constant independently of our choice of point P, we have

$$\ell_1 + \ell_2 = 2a. \tag{2.19}$$

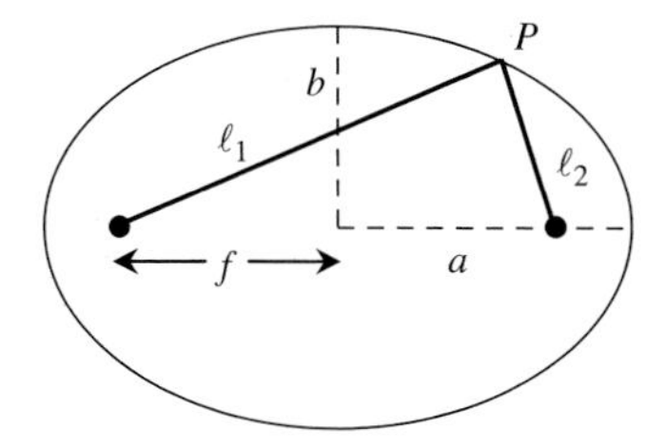

Fig. 2.10. The ellipse definition illustrated

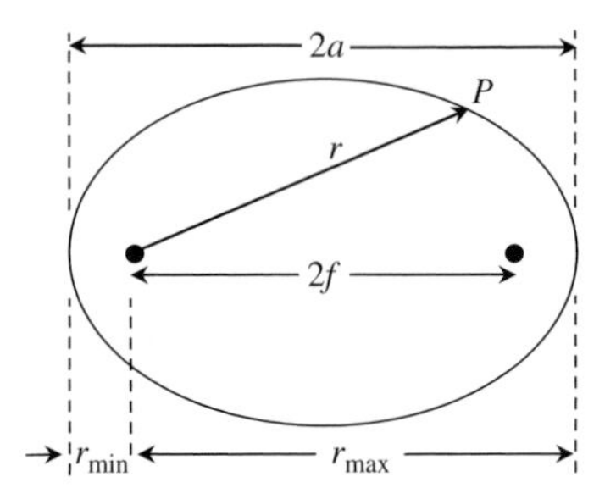

Fig. 2.11. Major axis: $2a$

Some other relationships which follow from Figures 2.10 and 2.11 are

$$r_{\text{min}} = a - f;$$

$$r_{\text{max}} = a + f \tag{2.20}$$

$$r_{\text{max}} + r_{\text{min}} = (a + f) + (a - f) = 2a \tag{2.21}$$

$$r_{\text{max}} - r_{\text{min}} = (a + f) - (a - f) = 2f \tag{2.22}$$

If we place point P at the end of the minor axis (Figure 2.13), then $\ell_1 = \ell_2$ and it follows from $\ell_1 + \ell_2 = 2a$ that the distance from either focus to the end of the minor axis is equal to the length of the semi-major axis, a.

Then, using Pythagoras' theorem in Figure 2.13, we have

$$b^2 = a^2 - f^2 = a^2 \left(1 - \frac{f^2}{a^2}\right) \tag{2.23}$$

We now define the eccentricity, e, of the ellipse as the ratio of the distance between the foci to the length of the major axis:

$$e = \frac{2f}{2a} = \frac{f}{a} \tag{2.24}$$

Substitution of (2.24) into (2.23) gives

$$b^2 = a^2 \left(1 - e^2\right) \tag{2.25}$$

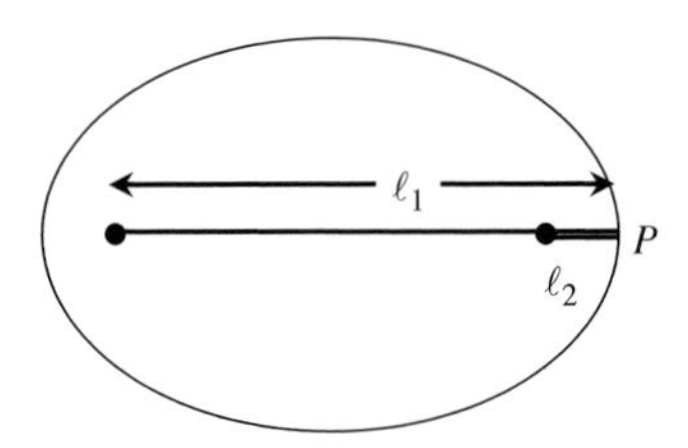

Fig. 2.12. Illustration of $\ell_1 + \ell_2 = 2a$

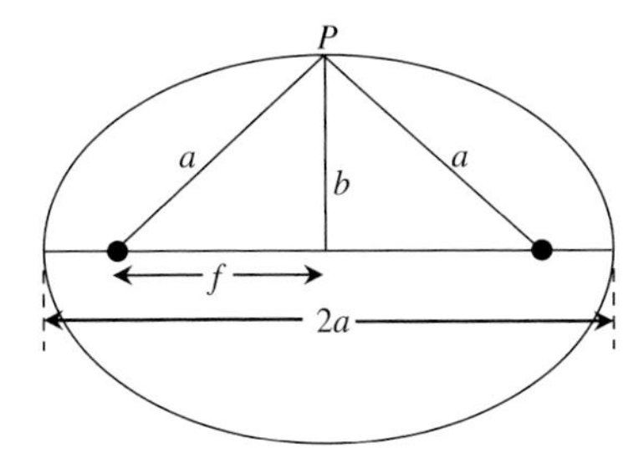

Fig. 2.13. The distance from either focus to one end of the semi-minor axis is equal to the length of the semi-major axis

Equations (2.20) and (2.25) then give

$$r_{\max} = a + f = a\left(1 + \frac{f}{a}\right) = a\,(1 + e) \tag{2.26}$$

$$r_{\min} = a - f = a\left(1 - \frac{f}{a}\right) = a\,(1 - e) \tag{2.27}$$

$$e = \frac{2f}{2a} = \frac{r_{\max} - r_{\min}}{r_{\max} + r_{\min}} \tag{2.28}$$

If we place the center of the ellipse at the origin of an (x,y) coordinate system as shown in Figure 2.14, then we can express the equation of the ellipse in terms of x and y as follows. First, from Figure 2.14 we have

$$\ell_1^2 = (f + x)^2 + y^2$$
$$\ell_2^2 = (f - x)^2 + y^2$$

Then using $\ell_1 + \ell_2 = 2a$ and $f^2 = a^2 - b^2$ (from equation (2.23)) and a lot of algebra gives the equation of an ellipse as

$$\frac{x^2}{a^2} + \frac{y^2}{b^2} = 1 \tag{2.29}$$

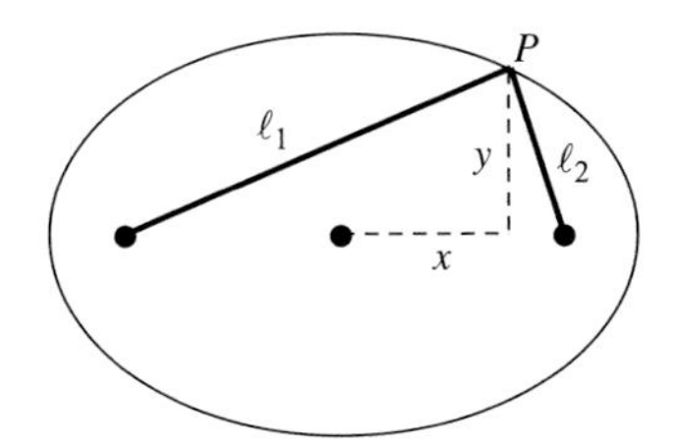

Fig. 2.14. x and y coordinates

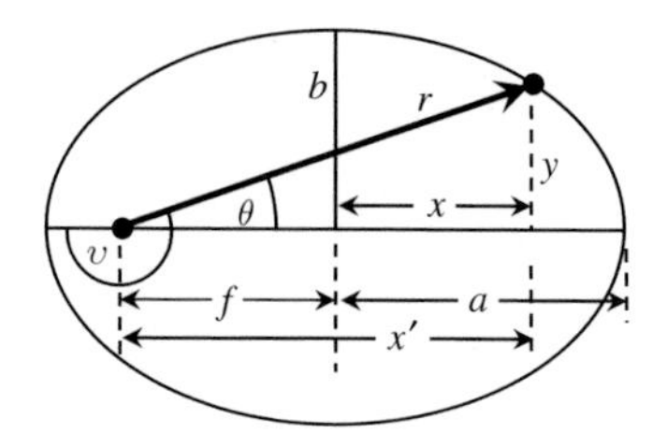

Fig. 2.15. Conversion from (x,y) to polar coordinates

We can also write the equation for an ellipse in polar coordinates centered on one focus, as follows. From Figure 2.15, substitute

$$x = x' - f = r\cos\theta - f$$
$$y = r\sin\theta$$

into (2.29), square the numerators, and rearrange terms to obtain the quadratic equation

$$r^2\left(\frac{\cos^2\theta}{a^2} + \frac{\sin^2\theta}{b^2}\right) - r\left(\frac{2f\cos\theta}{a^2}\right) + \left(\frac{f^2}{a^2} - 1\right) = 0 \tag{2.30}$$

The solution to (2.30) is

$$r = \frac{\frac{2f\cos\theta}{a^2} \pm \sqrt{\left(\frac{4f^2\cos^2\theta}{a^4}\right) - 4\left(\frac{\cos^2\theta}{a^2} + \frac{\sin^2\theta}{b^2}\right)\left(\frac{f^2}{a^2} - 1\right)}}{2\left(\frac{\cos^2\theta}{a^2} + \frac{\sin^2\theta}{b^2}\right)}$$
$$= \frac{f\cos\theta \pm a\sqrt{\cos^2\theta + \sin^2\theta\left(\frac{a^2-f^2}{b^2}\right)}}{\cos^2\theta + \frac{a^2}{b^2}\sin^2\theta}$$

Now $a^2 - f^2 = b^2$, so the argument of the square root reduces to 1 and r becomes, with the help of (2.28) and (2.25),

$$r = \frac{f\cos\theta \pm a}{\cos^2\theta + \frac{a^2}{b^2}\sin^2\theta} = \frac{f\cos\theta \pm a}{\cos^2\theta + \frac{\sin^2\theta}{1-e^2}} = \frac{a\,(e\cos\theta \pm 1)}{\left(\frac{1-e^2\cos^2\theta}{1-e^2}\right)}. \tag{2.31}$$

Regarding the $\pm$ sign, note that $e < 1$ and $\cos\theta \leq 1$. If we choose the negative sign then $(e\cos\theta - 1) < 0$ whereas the denominator is positive. This would make $r < 0$, which is impossible. Thus only the $+$ sign is relevant. Equation (2.31) then reduces to,

$$r = \frac{a\left(1 - e^2\right)}{1 - e\cos\theta} = \frac{r_{\min}\,(1 + e)}{1 - e\cos\theta}, \tag{2.32}$$

where $r_{\mathrm{min}} = a(1-e)$ from (2.27). In celestial mechanics the customary angle used is the *argument of perihelion*, angle υ in Figure 2.15; then $\cos \upsilon = \cos(\theta + 180°) = -\cos \theta$ and

$$r = \frac{a\left(1-e^2\right)}{1+e \cos \upsilon} = \frac{r_{\mathrm{min}}\,(1+e)}{1+e \cos \upsilon}. \tag{2.33}$$

The left equation of (2.33) describes an elliptical orbit. Kepler's 1st "law" is that planets move in elliptic orbits with the Sun at one focus. We now look at different limiting cases of the eccentricity. First, if we set $f=0$ then the two foci coincide at the center of the ellipse and $e = f/a = 0$. Then (2.23) and (2.25) both give $b = a$ (i.e., the semi-major and semi-minor axes are equal), and from (2.29) we have

$$\frac{x^2}{a^2} + \frac{y^2}{a^2} = 1 \quad \text{or} \quad x^2 + y^2 = a^2,$$

which is the equation for a circle. Thus a circle is an ellipse with an eccentricity of zero. If, on the other hand, we let $f \to \infty$ at constant r_{min}, then from (2.24) and (2.27),

$$e = \frac{f}{a} = \frac{f}{f + r_{\mathrm{min}}} \to 1.$$

The resulting curve is a parabola. If $e > 1$ then we have a hyperbola. These results are summarized in Table 2.1.

The area of an ellipse can be found by integration, as intimated in Figure 2.16:

$$A = 4\int_{x=0}^{a} \mathrm{d}A = 4\int_{x=0}^{a} y\,\mathrm{d}x \tag{2.34}$$

where, by equation (2.29), we have

$$y = b\left(1 - \frac{x^2}{a^2}\right)^{\frac{1}{2}} \tag{2.35}$$

Table 2.1. Eccentricities of ellipses and related curves

Curve	Eccentricity
Circle	0
Ellipse	$0 < e < 1$
Parabola	1
Hyperbola	> 1

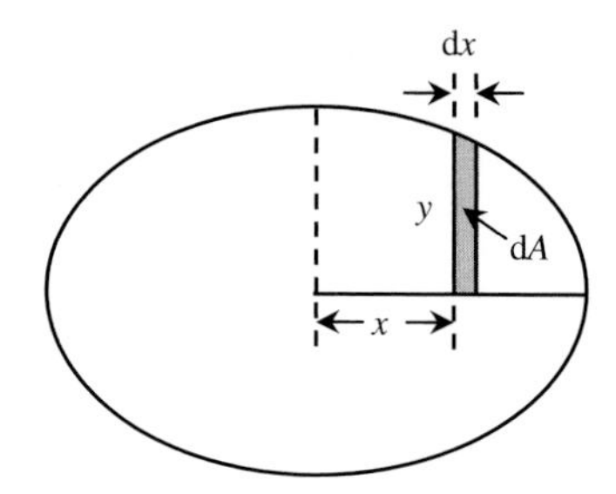

Fig. 2.16. Finding the area of an ellipse

This can be integrated with a trigonometric substitution,

$$x = a \sin \theta, \tag{2.36}$$

to obtain

$$A = \pi ab = \pi a^2 \sqrt{1 - e^2} \tag{2.37}$$

References

Green, R. M. 1985. *Spherical Astronomy* (Cambridge: Cambridge University Press).

Kelley, D. H. and Milone, E. F. 2005. *Exploring Ancient Skies: An Encyclopedic Survey of Archaeoastronomy* (New York: Springer).

Schlosser, W., Schmidt-Kaler, Th., and Milone, E. F. 1991/1994. *Challenges of Astronomy: Hands-On Experiments for the Sky and Laboratory* (New York: Springer).

Smart, W. M. 1977, *Spherical Astronomy* (Cambridge: Cambridge University Press). 6th or later ed., revised by R. M. Green.

Challenges

[2.1] Compute the great circle distance and the initial and final bearings (= azimuths) for a voyage to Singapore (λ = 103.85°E, ϕ = 1.28°N) from Vancouver (λ = 123.2°W, ϕ = 49.3°N).

[2.2] Write down the equations of transformation between the equatorial and ecliptic systems beginning with a drawing of the appropriate spherical triangle.

[2.3] Compute the celestial longitude and latitude of an object at $\alpha = 15^{\mathrm{h}}\ 39^{\mathrm{m}}\ 40^{\mathrm{s}}$, $\delta = -5^\circ\ 17.8'$.

[2.4] Derive the equations of transformation between the horizon and (H,δ) equatorial system.

[2.5] For a site with latitude $\lambda = 38°$ N, what are the equatorial coordinates of:

(a) a star located at the NCP?

(b) a star on the horizon at the South point at 12^{h} local sidereal time?

(c) a star overhead at local midnight on March 21?

[2.6] For a site with latitude $\lambda = 38°$S, what are the equatorial coordinates of:

(a) a star located at the SCP?

(b) a star on the horizon at the North point at 12^{h} local sidereal time?

(c) a star overhead at local midnight on March 21?

3. Celestial Mechanics

An important field for solar system studies is the area of physics of motion, that is, kinematics and dynamics, as applied to celestial objects. This field is called celestial mechanics. It is impossible to do full justice to this subject within a single chapter, but we will provide an introduction that may be useful for further studies and provide a background to understanding planetary dynamics as we apply it in this book. We begin first with an understanding of the orbits of the planets and of the other objects which move about the Sun.

3.1 The Two-Body Problem

The geometry of the solar system can be considered on many scales. On the largest scales, the potential well created by the Sun dominates the scene, on the smallest gravitational scales (that we have discerned), the mutual interactions of moons and ring fragments are noticeable (see Chapter 13 of Milone & Wilson, 2008). The basic equations in celestial mechanics are Newton's gravitational law and laws of motion.

Given two objects of mass m_1 and m_2, we want to determine the orbit of either one around the other, assuming that the only force acting is their mutual gravitational attraction. In the process, we will derive two of Kepler's three laws as well as a number of other useful quantities and equations.

Figure 3.1 shows an inertial x,y reference frame with origin O and masses m_1 and m_2 at positions marked by the vectors $\mathbf{r}_1$ and $\mathbf{r}_2$, respectively. We take the vector $\mathbf{r}$ as pointing from m_1 to m_2, and the unit vector, $\hat{\mathbf{r}}$, as pointing in the same direction (i.e., radially out from m_1 to m_2). Note that

$$\mathbf{r} = r\hat{\mathbf{r}}$$

where the quantity r is a scalar, having size or magnitude, but containing no direction information.

According to Newton's third law, the gravitational forces on each of the two objects are equal in magnitude and opposite in direction, and according to

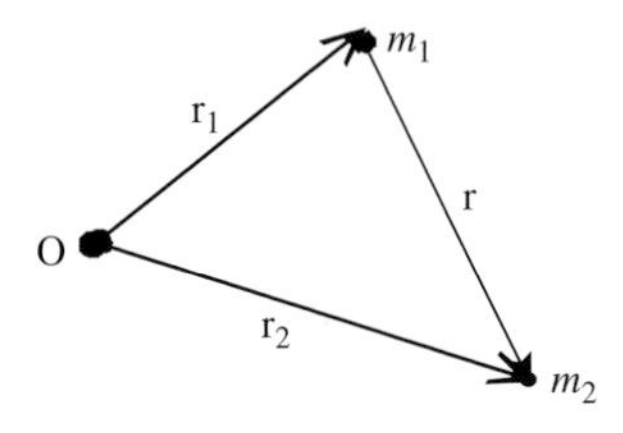

Fig. 3.1. Two masses in an inertial coordinate system

Newton's law of gravitation the magnitude of each force is

$$\mathbf{F}_{\mathrm{g}} = \frac{Gm_1m_2}{r^2} \tag{3.1}$$

Then by Newton's second law the force on m_1 is

$$\mathbf{F}_1 = m_1\mathbf{a}_1 \equiv m_1\ddot{\mathbf{r}}_1 = \frac{Gm_1m_2}{r^2}\hat{\mathbf{r}} \tag{3.2}$$

Then the equation of motion of m_1 in the x,y frame shown in Figure 3.1 is given by

$$\ddot{\mathbf{r}}_1 = \frac{Gm_2}{r^2}\hat{\mathbf{r}} \tag{3.3}$$

Here we have shown time derivatives by dots over the quantity, so, for example, the acceleration, a, of m, can be written as "r double dot." Formally,

$$\dot{\mathbf{r}} \equiv \frac{\mathrm{d}\mathbf{r}}{\mathrm{d}t}; \qquad \ddot{\mathbf{r}} \equiv \frac{\mathrm{d}^2\mathbf{r}}{\mathrm{d}t^2}$$

Similarly, the equation of motion of m_2 is obtained from

$$\mathbf{F}_2 = m_2\mathbf{a}_2 \equiv m_2\ddot{\mathbf{r}}_2 = \frac{Gm_1m_2}{r^2}(-\hat{\mathbf{r}}) \tag{3.4}$$

where $\mathbf{F}_2$ is in the negative or opposite direction to the vector $\mathbf{r}$. Then,

$$\ddot{\mathbf{r}}_2 = -\frac{Gm_1}{r^2}\hat{\mathbf{r}} \tag{3.5}$$

Now $\mathbf{r} = \mathbf{r}_2 - \mathbf{r}_1$ as shown in Figure 3.1. Differentiating twice with respect to time gives

$$\ddot{\mathbf{r}} = \ddot{\mathbf{r}}_2 - \ddot{\mathbf{r}}_1 \tag{3.6}$$

so that

$$\ddot{\mathbf{r}} = \left(-\frac{Gm_1}{r^2}\hat{\mathbf{r}}\right) - \left(+\frac{Gm_2}{r^2}\hat{\mathbf{r}}\right) \qquad \text{or}$$

$$\ddot{\mathbf{r}} = -\frac{G(m_1+m_2)}{r^2}\hat{\mathbf{r}} \tag{3.7}$$

The $(-)$ sign in this equation shows that the acceleration acts to shorten r, that is, gravity is an attractive force. This is Newton's gravitational law for relative orbits.

With equation (3.7), we can forget about $\mathbf{r}_1$ and $\mathbf{r}_2$ and work directly with $\mathbf{r}$. Consider the general expression for the force exerted on a planet of mass m, by the Sun, with mass $\mathfrak{M}_\odot$:

$$\mathbf{F} = -[G\mathfrak{M}_\odot \mathrm{m}/r^2]\,\hat{\mathbf{r}} = -[G\mathfrak{M}_\odot \mathrm{m}/r^3]\mathbf{r} \tag{3.8}$$

where the vector $\mathbf{r} = r\hat{\mathbf{r}}$ is in the direction of the planet (of mass m), away from the Sun and G is the gravitational constant with the determined value $6.67259 \pm 0.00085 \times 10^{-11}$ $\mathrm{m^3/kg\,s^2}$ (Cohen and Giacomo 1987, Table 9). Here, again, the negative sign indicates an attraction toward the Sun. Newton's second law of motion then permits us to arrive at the critical differential equation:

$$m\ddot{\mathbf{r}} = -[G\mathfrak{M}_\odot m/r^3]\mathbf{r} \tag{3.9}$$

In (3.9), we have moved the origin in (3.7) to the center of the more massive body, in the solar system context, the Sun. In the relative orbit, (3.9) effectively ignores the mass of the lesser object (i.e., $m << \mathfrak{M}_\odot$), so that $\mathfrak{M}_\odot + m \approx \mathfrak{M}_\odot$, a condition that holds to one part in 1000 or better in the solar system.

In spherical coordinates (r, the distance from the Sun, and θ, the angular variable), one may express the velocity as: $\dot{\mathbf{r}} = \dot{r}\hat{\mathbf{r}} + r\dot{\theta}\;\hat{\theta}$, and making use of the conventions D $\hat{\mathbf{r}} = \dot{\theta}\;\hat{\theta}$, and D $\hat{\theta} = -\dot{\theta}\;\hat{\mathbf{r}}$, where $\hat{\theta}$ is the coordinate direction $\perp$ to $\hat{\mathbf{r}}$, and D is the time-derivative operator, d/dt, we get for the relative acceleration

$$\ddot{\mathbf{r}} = (\ddot{r} - r\dot{\theta}^2)\hat{\mathbf{r}} + (r\ddot{\theta} + 2\dot{r}\dot{\theta})\,\hat{\theta} \tag{3.10}$$

By definition there is no $\hat{\theta}$ component for a centrally directed force, so $r\ddot{\theta} + 2\dot{r}\dot{\theta} = 0$. However, $r\ddot{\theta} + 2\dot{r}\dot{\theta} = (1/r)$ D$(r^2\dot{\theta})$ so D$(r^2\dot{\theta}) = 0$, and therefore

$$r^2\dot{\theta} = \text{constant} \tag{3.11}$$

The *constant* is usually called h, the angular momentum (per unit mass). Because

$$r^2\dot{\theta} = \mid \mathbf{r} \times \dot{\mathbf{r}} \mid = r\dot{r}\sin\;\phi \tag{3.11a}$$

where ϕ is the angle between the vectors $\mathbf{r}$ and $\dot{\mathbf{r}}$, it can be seen that h is actually a vector and is $\perp$ to the plane containing $\mathbf{r}$ and $\dot{\mathbf{r}}$. The constancy of $\mathbf{h}$ in magnitude and in direction assures that the planet continues to move only in the original plane.

It may be also shown that the rate of change of the area swept out by $\mathbf{r}$, the areal velocity, is

$$\Delta A/\Delta t = r^2\dot{\theta}/2 = h/2 \tag{3.12}$$

which is a succinct statement of Kepler's second law.

To arrive at the energy equation for an orbit, form the dot product of the vectors $\dot{\mathbf{r}}$ and $\ddot{\mathbf{r}}$:

$$\dot{\mathbf{r}} \cdot \ddot{\mathbf{r}} = 1/2\ D(\dot{\mathbf{r}} \cdot \dot{\mathbf{r}}) = 1/2\ D[\dot{r}^2 + (\dot{\theta}\ r)^2] \tag{3.13}$$

and also, from (3.7), writing $\mu = G(\mathfrak{M}_\odot + m)$,

$$\dot{\mathbf{r}} \cdot \ddot{\mathbf{r}} = -\mu\ \dot{\mathbf{r}} \cdot \mathbf{r}/r^3 = -\mu\ \dot{r}/r^2 = \mathrm{D}(\mu/r) \tag{3.14}$$

Combining the integrands of equations (3.13) and (3.14), we obtain the energy equation:

$$(1/2)(\dot{r}^2 + r^2\dot{\theta}^2) - (\mu/r) = \text{constant} \tag{3.15}$$

where the constant is the total energy per unit mass, E. With (3.11), (3.15) becomes

$$\dot{r}^2 = 2(E + \mu/r) - h^2/r^2 \tag{3.16}$$

Making use of the equation

$$\mathrm{d}r/\mathrm{d}t = (\mathrm{d}r/\mathrm{d}\theta)(\mathrm{d}\theta/\mathrm{d}t) \tag{3.17}$$

and after some manipulation, we get

$$(\mathrm{d}\theta/\mathrm{d}r) = (h/r^2)/\sqrt{[2(E + \mu/r) - h^2/r^2]} \tag{3.18}$$

After a change in variable, $u = 1/r$, and some manipulation, (3.18) becomes, in integral form,

$$\theta = -\int(a - bu + u^2)^{-1/2}\mathrm{d}u \tag{3.19}$$

where constants $a \equiv 2E/h^2$ and $b \equiv 2\mu/h^2$; N.B.: this a is not the semi-major axis of Chapter 2.2.

The solution, by quadratures, is found in tables of integrals, such as in Peirce (1957), where it is No. 166:

$$\int \mathrm{d}x\ (X)^{-1/2} = -(-c)^{-1/2} \arcsin[(2cx + b)(-q)^{-1/2}] \tag{3.20}$$

where $X = a + bx + cx^2$ and $q = 4ac - b^2$, so that the solution is

$$\theta = \arcsin\left[-(2u + b)/(4a + b^2)^{-1/2}\right] + \gamma \tag{3.21}$$

and γ is a constant of integration. From this,

$$u = -(\mu/h^2)\{1 + [2Eh^2/\mu^2 + 1]^{1/2}\ \sin\ (\theta - \gamma)\} \tag{3.22}$$

We take γ to be 90° so that

$$r = (h^2/\mu)/\{\surd[(2Eh^2/\mu^2) + 1]\ \cos\ \theta + 1\} \tag{3.23}$$

Setting

$$\begin{aligned} h^2/\mu = a(1 - e^2) = q(1 + e) \text{ and} \\ \surd[(2Eh^2/\mu^2) + 1] = e \end{aligned} \tag{3.24}$$

where $q = a(1-e)$ is the perihelion distance [see equations (2.20) and (2.32)], a is the semi-major axis, and e is the eccentricity, (3.24) becomes

$$r = q(1 + e)/[e\ \cos\ \theta + 1] = q(1 + e)/[1 + e\ \cos\ \theta] \tag{3.25}$$

This form is applicable to all orbits, i.e., with all values of eccentricity, but ellipses are much more commonly encountered in the solar system.

3.2 Orbital Elements

That planets move in ellipses, is the statement of Kepler's first law. The properties of the orbit can be summarized by the constants of the integration. The basic differential equation (3.10) is a second-order equation, so there are six constants of integration. We have discussed the quantities h and E, which are related to the geometrical quantities a and e. The geometrical orbital elements are:

a the *semi-major axis* of the ellipse, which establishes the scale of the orbit

e the *eccentricity*, which establishes the shape

i the *inclination*, which establishes the tilt of the orbital plane, relative to the ecliptic plane

Ω the *longitude of the ascending node* (ascending crossover point of the orbit, measured with respect to the line to the vernal equinox), which establishes the orientation of the orbital plane

ω the *argument of perihelion* (measured from the ascending node to the perihelion point; the sense is in the direction of orbital motion), which establishes the orientation of the ellipse within the orbital plane

T_0 the *epoch*, which in this context, is an instant when the object is at perihelion

The elements are illustrated in Figure 3.2. If the inclination is very small so that the ascending node is not well-established, and/or the inclination is very close to zero, the *longitude of perihelion*, ϖ ("curly pi") $= \Omega + \omega$ may be used in place of Ω and ω. The *period* of revolution, P, is sometimes included in the elements; it is not quite redundant, even though Kepler's third law ($\mu P^2 = 4\pi^2 a^3$), would suggest that it is [recall that the

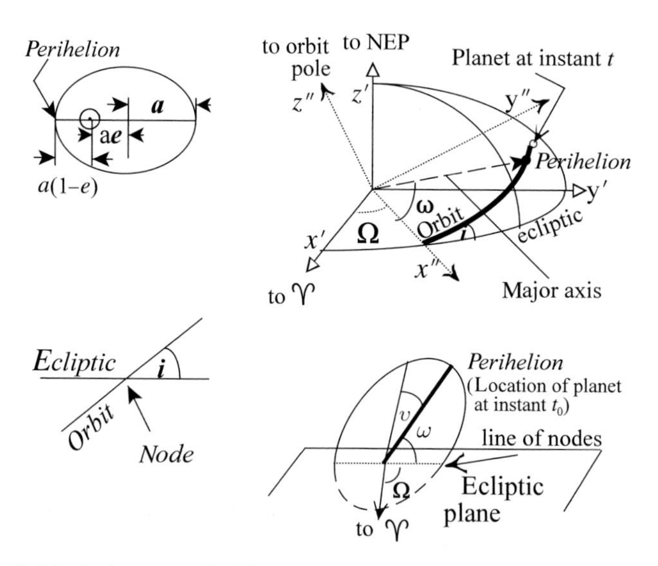

Fig. 3.2. Orbital elements (*bold*) and other quantities, in plan, elevation, and oblique perspectives of the planetary orbit and the ecliptic

quantity $\mu = G(\mathfrak{M}_{\odot} + m)$ really depends on the mass of the planet as well as that of the Sun].

In the computation of ephemerides (predictions of position of the planet with time), the angular variable (θ), or more specifically the *true anomaly*, (υ), identical to θ in equations (3.23) and (3.25), is computed from more directly time-related quantities, the *eccentric anomaly* and the *mean anomaly*. Kepler's equation, discussed in Section 3.5, relates these quantities. In Section 3.4 we describe the computation of ephemerides given the orbital elements.

In general, nature is more complicated than our self-consistent discussion of orbits would lead one to believe. Although the motions of two isolated bodies can be solved analytically and therefore exactly, this is in general not true for more than two bodies. The reason: mutual gravitational interactions make exact solutions impossible. However, numerical methods and high-speed computation have made dramatic progress in n-body analyses and computation of ephemerides. In Section 3.8, we discuss the evaluation of perturbations to orbital motions. Moreover, there are exceptions to the lack of closed analytic solutions. In the solar system, for instance, the interaction of some asteroids with Jupiter and the Sun illustrates a famous condition: *the restricted three-body problem.* The condition is also approximated in the satellite systems of the giant planets. Therefore, we describe this condition next.

3.3 The Restricted Three-Body System

Under certain circumstances in a three-body system, a close approximation to an exact solution is possible. If one of the bodies has an infinitesimal and therefore negligible mass, and the orbits of the two more massive bodies are circular, then an analytic solution can be found. Because real orbits will not be exactly circular, and no real body can be completely massless, only an approximation to this condition can be achieved. But the masses of the Sun and Jupiter are so large relative to the small bodies of the solar system, and the eccentricity of Jupiter's orbit sufficiently small, that the approximation is valid to rather high precision. The solutions include several points in a frame of reference that co-rotates with the two massive objects. At these *Lagrangian points*, a third body (or group of small bodies) can, once there, remain. These are called $L_1, \ldots, L_5$, depending on their locations with respect to the greater and smaller of the two large masses. As illustrated in Figure 3.3, created with the aid of Binary Maker 3 (Bradstreet and Steelman 2004), L_1 is between the two massive objects; the colinear points L_2 and L_3 are on either side, and all three are on the line joining the

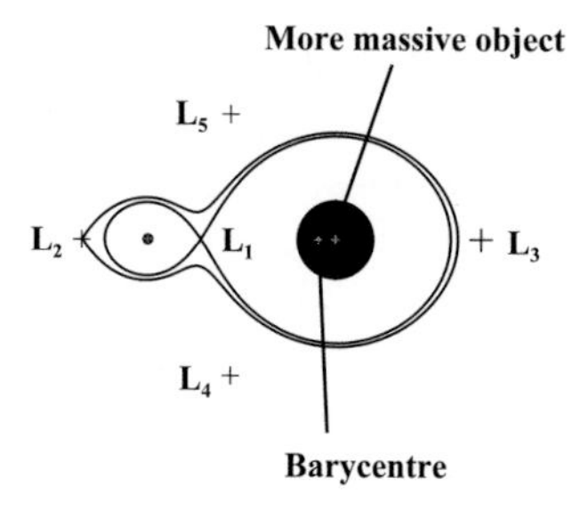

Fig. 3.3. The Lagrangian points: solutions of the restricted three-body problem. Illustration created with the help of Binary Maker 3 (Bradstreet and Steelman 2004)

centers of the massive pair of objects. L_4 and L_5 are perpendicular to this line and form equilateral triangles with the massive objects. The first three are known to be quasi-stable points: small objects will remain if unperturbed; the last two are stable points, where objects can resist all but the largest perturbations.

To return to our example, the Greek and Trojan asteroids continuously chase each other around the solar system while Zeus and Apollo dominate the scene and witness the events.

Orbital parameters of planetary orbits vary with time, mostly in cyclic ways. The motions of the smaller bodies are generally more complex and may be unpredictable (aspects of Pluto's motion can be characterized this way), for reasons that we discuss later. For an instant of time, the orbital elements characterizing a particular orbit may be specified: they are the "osculating" orbit elements, i.e., they "kiss" the true orbit at the epoch for which the elements are specified. Beyond this instant, each element is subject to perturbations.

3.4 Computation of Ephemerides

If an instant of time is specified, then, given the orbital elements, the position in the orbit, and in the sky, can be obtained as follows.

1. Compute first the *mean anomaly* M from its defining equation:

$$M \equiv 2\pi(t - T_0)/P \qquad (3.26)$$

where t and T_0 are expressed in Julian Day numbers (JDN), typically. Note that M is an angular variable that varies linearly with time, so that if the orbit were circular, M would be identical with the true anomaly, υ. Usually, $e \neq 0$; but in this case, depending on the size of the eccentricity, υ can be found either (a) from an approximation in terms of M directly or (b) more exactly but less directly, with the help of the eccentric anomaly, E, which we define and describe in Section 3.5, below. For now, we take route (a); therefore, having M, we find υ as follows.

2. If the eccentricity is sufficiently small (if not, see the next section!), we may use the approximation

$$\upsilon = M + 2e \sin M + (5/4)e^2 \sin 2M + (1/12)e^3(13 \sin 3M - 3 \sin M) + (1/96)e^4(103 \sin 4M - 44 \sin 2M) + \cdots \tag{3.27}$$

Then, with υ in hand, with equation (3.25), and substituting $q = a(1 - e)$,

3. Compute the radius vector:

$$r = a(1 - e)^2/(1 + e \cos \upsilon) \tag{3.28}$$

The rectangular (right hand system) coordinates in the orbital plane are then $x_0 = r \cos \upsilon$, $y_0 = r \sin \upsilon$, $z_0 = 0$, where the x-axis is in the direction of perihelion, and the y-axis, 90° ahead in the plane of the orbit. This gives the position of the planet within the orbit.

As an aside, in addition to the variable υ, the true or mean longitudes are sometimes used to describe the location of an object in the orbital plane. These are defined as follows:

- *True longitude*, $L = \Omega + \omega + \upsilon$. In turn the *longitude of perihelion* (more generally, *pericenter*), $\varpi = \Omega + \omega$ is sometimes used, especially when the node is ill-defined (i.e., when the inclination is near zero), and the *argument of latitude*, $u = \omega + \upsilon$ is sometimes used, especially when the pericenter is ill-defined because e is very close to zero.

  Thus, $L = \varpi + \upsilon$ and $L = \Omega + u$ are equivalent.
- *Mean longitude*, $\ell = \varpi + n(t - T_0) = \varpi + M$

  where n is the mean motion, formally defined by (3.39) below, and M is the mean anomaly.

4. To specify where the planet is in the sky, we can first calculate the position in the ecliptic and then transform this into the equatorial system (see Chapter 2.1.4) or we may make use of a series of auxiliary relations, first introduced by Karl Friedrich Gauss (1777–1855) to go more directly into the equatorial system (see Moulton 1914, p. 188):

$$\begin{aligned}
\alpha' \sin A &= \cos\ \Omega \\
\alpha' \cos &= -\sin\Omega\ \cos\ i \\
\beta' \sin B &= \sin\Omega\ \sin\ \epsilon \\
\beta' \cos B &= \cos\Omega\ \cos\ i\ \cos\ \epsilon \\
\gamma' \sin C &= \sin\Omega\ \sin\epsilon \\
\gamma' \cos C &= \cos\Omega\ \cos i\ \sin\ \epsilon
\end{aligned} \tag{3.29}$$

where Ω and i are the orbital elements, as defined earlier, ϵ is the *obliquity of the ecliptic*, and the quantities α', β', γ' (which Gauss called sin a, sin b, and sin c, respectively), and A, B, and C, need to be computed only once per orbit. From these equations, the heliocentric equatorial coordinates become

$$\begin{aligned}
x &= r\alpha' \sin\ (A + \omega + v) \\
y &= r\beta' \sin\ (B + \omega + v) \\
z &= r\gamma' \sin\ (C + \omega + v)
\end{aligned} \tag{3.30}$$

where ω is the argument of perihelion. Finally, the geocentric equatorial coordinates may be computed:

$$\begin{aligned}
\xi &= x + X \\
\eta &= y + Y \\
\zeta &= z + Z
\end{aligned} \tag{3.31}$$

where X, Y, and Z are the rectangular coordinates of the Sun in geocentric equatorial coordinates, tabulated annually in the Astronomical Almanac as quantities labeled "x," "y," and "z," respectively. The relationship between these quantities and right ascension and declination is given by

$$\begin{aligned}
\rho\ \cos\ \delta\ \cos\ \alpha &= \xi \\
\rho\ \cos\ \delta\ \sin\ \alpha &= \eta \\
\rho\ \sin\ \delta &= \zeta
\end{aligned} \tag{3.32}$$

where ρ is the geocentric distance of the planet:

$$\rho = [\xi^2 + \eta^2 + \zeta^2]^{1/2} \tag{3.33}$$

so,

$$\sin\delta = \zeta/\rho \quad \text{and} \quad \tan\alpha = \eta/\xi \tag{3.34}$$

More detailed treatment can be found in Moulton (1914, pp. 182–189) or in other texts on Celestial Mechanics [e.g., Brouwer and Clemence (1961), Danby (1962/1964; 1988), Neutsch and Scherer (1992), Smart (1953), or Szebehely (1989)], and, in somewhat more abridged but focused discussions, in Montenbruck (1989) or Schlosser et al. (1991/4).

3.5 Kepler's Equation

A useful starting point is the *vis-viva* equation, which may be derived from the energy equation (3.15). At perihelion, the energy equation may be written as

$$E = (1/2)(\dot{r}^2 + r^2\dot{\theta}^2) - (\mu/r) = (1/2)(0 + r^2\dot{\theta}^2) - (\mu/r) \tag{3.35}$$

and because $r^2\dot{\theta}^2 = h^2/r^2$, recalling [see equation (3.24) and the text near it] that $h^2/\mu = a(1-e^2)$, and that at perihelion, $r = a(1-e)$, this simplifies to

$$E = -(1/2)\mu/a \tag{3.36}$$

E is a constant, so it has the same value everywhere in the orbit. Substituting and setting v^2 equal to the sum of the squared velocity terms in (3.15), we get the *vis-viva* equation:

$$v^2 = \mu\ [(2/r) - (1/a)] \tag{3.37}$$

If we now make use of Kepler's third law, sometimes referred to as "Kepler III,"

$$a^3 = [\mu/(4\pi^2)]P^2 \tag{3.38}$$

and define the *mean* (angular) *motion*, n, by

$$n \equiv 2\pi/P \tag{3.39}$$

the latter becomes

$$n = \mu^{1/2}/a^{3/2} \tag{3.40}$$

Because, in general,

$$v^2 = \dot{r}^2 + (r\,\dot{\theta})^2$$
$$(r\,\dot{\theta})^2 = h^2/r^2$$

and

$$h^2/\mu = a(1 - e^2)$$

we have,

$$v^2 = \dot{r}^2 + h^2/r^2 = \dot{r}^2 + \mu(a/r^2)(1 - e^2) \tag{3.41}$$

Also, by squaring (3.40), and rearranging, we get

$$\mu = n^2 a^3 \tag{3.42}$$

and, with this substitution into (3.37),

$$v^2 = n^2 a^3 (2a - r)/(ar) \tag{3.43}$$

Then, solving for the square of the derivative of the distance scalar on the RHS of (3.41), and substituting for μ from (3.42) and for v^2 from (3.43), we obtain

$$\dot{r}^2 = [n^2 a^2 (2a - r) r - n^2 a^4 (1 - e^2)]/r^2 \tag{3.44}$$

From this,

$$\begin{aligned} n\,\mathrm{d}t &= (r/a)[2ar - r^2 - a^2(1 - e^2)]^{-1/2}\,\mathrm{d}r \\ &= (r/a)[a^2 e^2 - (a - r)^2]^{-1/2}\,\mathrm{d}r \end{aligned} \tag{3.45}$$

We now define the *eccentric anomaly,* traditionally written as E (but not to be confused with the orbital energy), such that

$$r = a(1 - e\cos E) \tag{3.46}$$

From this, $(a - r)$, r/a, and the differential $\mathrm{d}r = ae\,\sin E\,\mathrm{d}E$ are obtained, and after substitution in (3.45),

$$n\,\mathrm{d}t = (1 - e\cos E)\,\mathrm{d}E \tag{3.47}$$

which, when integrated, yields

$$n(t - T_0) = E - e\sin E \tag{3.48}$$

where T_0 is the initial value of time (an instant when $E = 0$, i.e., at pericenter), and t the terminal time, the instant of interest.

Then the *mean anomaly*, M [defined in (3.26)], may be written as

$$M = n(t - T_0) \tag{3.49}$$

With (3.48), *Kepler's equation* emerges:

$$M = E - e \sin E \tag{3.50}$$

Figure 3.4 illustrates how the eccentric anomaly, E, is related to the true anomaly, υ. Note that neither moves at a linear rate. Like E, M is measured from the center, but it increases at a constant rate. The geometry of this figure may be used to derive Kepler's equation geometrically, and we now proceed to do so.

The geometrical interpretation of Kepler's equation (note that *Kepler's equation* should not be confused with any of Kepler's three "laws") may be demonstrated as follows.

Because M is an angle that increases uniformly with time, and because the areal velocity is constant [refer to the text near equations (3.11) and (3.12)],

$$M/2\pi = \text{area P} \odot \text{R/ellipse area} = \text{area T} \odot \text{P/circle area}$$

but

$$\text{area T} \odot \text{P} = \text{area TCP} - \text{area TC}\odot$$

The area of the segment of the circle is

$$\text{area TCP} = (1/2)a(aE)$$

where the angular quantity E, when treated algebraically outside of trigonometric functions, as here, must be expressed in radian measure. The line segment $\text{C}\odot = ae$, so the triangular area

$$\text{TC}\odot = (1/2)(ae)\ \text{TQ} = (1/2)(ae)\ a\ \sin\ E$$

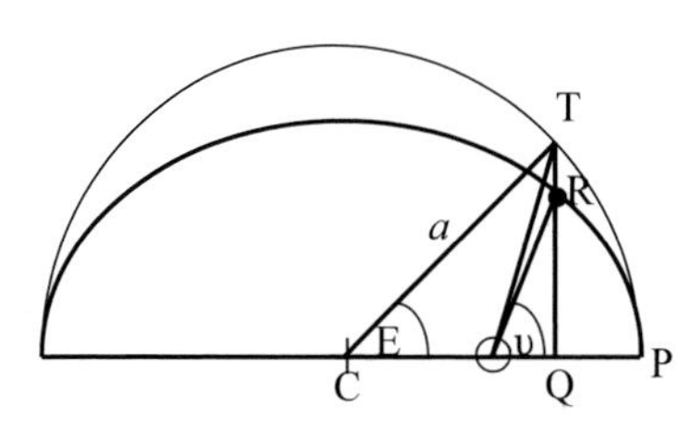

Fig. 3.4. The relationship between υ and E. Note that TP is a circular arc; RP is an elliptical arc

Therefore,

$$\text{area T} \odot \text{P} = (E/2)a^2 - (1/2)(ae)\ a\ \sin\ E$$

whence,

$$(M/2\pi) = (a^2/2)(E - e\ \sin\ E)/(\pi a^2)$$

or

$$M = E - e\ \sin\ E$$

which is Kepler's equation, once again. The study of this equation has a long and venerable history. See Neutsch and Scherer (1992), especially Chapters 3 and 4, for both history and method.

It is necessary to solve Kepler's equation for E in order to obtain the position of the object at the instant t. The relation between υ and E is obtained by setting equal equations (3.28) and (3.46):

$$a(1 - e^2)/(1 + e\ \cos\ \upsilon) = a(1 - e\cos E)$$

whence,

$$\cos\upsilon = (\cos E - e)/(1 - e\cos E) \tag{3.51}$$

There are several ways to achieve solutions for Kepler's equation, some more easily, others more precisely. Some of these are:

1. Graphically, one can plot the functions $f = \sin E$ and $g = (1/e)(E - M)$. The intersection of the two curves is the solution.
2. Tabularly, one can examine, perhaps interpolate, among these quantities to find the solution.
3. Iteratively, which is best suited for computation. We illustrate:

   (a) Start with a rough solution, say E_0.

   (b) Compute

   $$M_0 = E_0 - e\ \sin E_0 \tag{3.52a}$$

   (c) But M is known, so compute the difference

   $$\Delta M = M - M_0 \tag{3.52b}$$

(d) Then, differencing (3.50),

$$\Delta M = \Delta E - (e \cos E)\, \Delta E \tag{3.52c}$$

compute the correction,

$$\Delta E_0 = \Delta M / (1 - e \cos E) \tag{3.52d}$$

(e) Then find a new value for E:

$$E_1 = E_0 + \Delta E_0 \tag{3.52e}$$

(f) Increase the subscripts by one, and repeat steps (b), through (e), until ΔM matches the uncertainty in M.

Examples can be found in Danby (1962, p. 148ff; 1988, p. 149ff) and in Moulton (1914, p. 160ff, 181 (#3)).

It may be shown further that E is expressible as an expansion of M, resulting in a series. The expression recapitulates the iterative procedure to a given order.

3.6 Uses and Limitations of Two and Three-Body Solutions

First we illustrate a use for the two-body system, one that has generated billions of dollars of profit for communications satellite corporations.

Example 3.1 Use of two-body system celestial mechanics

The two-body equations of motion suffice for approximate solutions in the presence of a dominating mass and in the absence of major sources of perturbations. For example, consider the case of a synchronous satellite. It is possible to launch a small satellite with a rocket into low Earth orbit and then to transfer it to another, higher orbit. Higher orbits have the advantage of little atmosphere drag on the satellite, which causes a loss of orbital energy, and eventual orbital decay. A particular type of high orbit has another great advantage for communications: a synchronous satellite (at present found only around the Earth, where it is called a *geo-synchronous satellite*) orbit, illustrated in Figure 3.5.

We have already mentioned that two-body systems, although analytically rigorous to describe, are not sufficient to describe the mechanics of objects moving in the solar system to the highest precision, in most cases.

The *restricted three-body* problem has a closed or analytical solution. This is the only exact solution for the interaction of three bodies and it requires special circumstances: one of the three bodies must have such a sufficiently small mass that it has no significant effect on the other two bodies.

We mentioned in section 3.3 the solutions for the restricted three-body problem and refer to them again later, but it is not the only possible way to solve multi-body interactions. Numerical calculations can be carried out and, in this way, orbits can be plotted and object positions predicted, for the major bodies of the solar system. This is how space mission trajectories are computed, for example, even in the presence of major perturbation sources.

The basic idea for n-body orbital calculations is this:

1. The distances of the object in question from all significant masses are established for some instant, from the osculating orbit of the object perhaps, and the theories of motion of all the other bodies.
2. Then the force due to each of the objects is computed to find the total force, and the net acceleration of the body found.

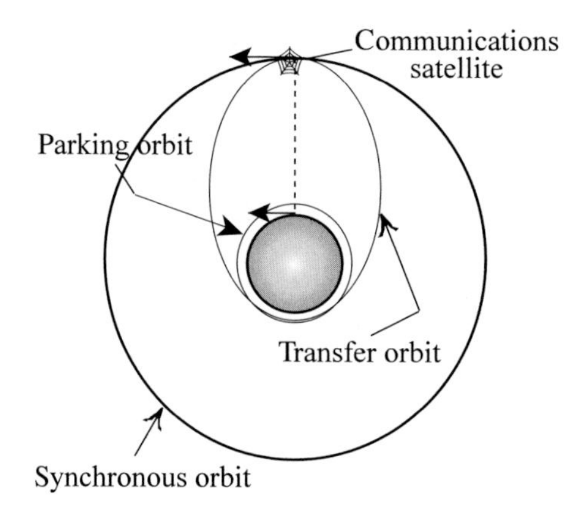

Fig. 3.5. The synchronous satellite and transfer orbits. An object in a geo-synchronous orbit has the same period, and thus angular speed, as an object on the Earth's surface, subjected to only the Earth's rotation

3. The acceleration (multiplied by a small time step) is then applied as a vector addition to the previous velocity of the body.
4. The mean velocity over the time step (a small interval) is calculated and the net change in position computed; return to stage (1).

See Brouwer and Clemence (1961, p. 171ff) for an example, the procedure for computing the ephemeris of the asteroid *1 Ceres*.

Modern computers have greatly aided the process of integration that is involved in predicting positions accurately. One modern system is the *Digital Orrery* (Applegate et al., 1985). As with most intensive and repeated operations, the limiting precision at each step determines the accuracy of the end result.

A major difficulty in multi-body orbital predictions is the presence of families of orbits where chaos can cause unpredictability. A circumstance can arise wherein a very small change in the force results in a significantly different motion. Some asteroid, cometary, and satellite orbits, and even those of "dwarf planets" (e.g., Pluto) are subject to this. Because of this, in some cases we cannot really be sure what the remote past was like, and we can be even less secure about the future. The past at least has the constraint that the endpoint of the orbital evolution is the present configuration; in the worst cases, we may not know what the configuration will be in the distant future.

Now, however, we consider an interesting problem: the transfer of angular momentum from orbital motion into rotational motion, and vice versa.

3.7 Spin–Orbit Coupling

The coupling between rotation and orbit requires some interaction other than the inverse square law operating between two effective mass centers. The most common such interaction is tidal.

3.7.1 Effect of Tidal Friction on Rotation

The tidal interaction between two objects with center-to-center distance r can be expressed as

$$\begin{aligned}\Delta F &= -\,G\mathfrak{M}m[(r-\Delta r)^{-2}-r^{-2}]\\ &= -\,G\mathfrak{M}m[2r\Delta r-(\Delta r)^2]r^{-2}\,[r^2-2r\Delta r+(\Delta r)^2]^{-1}\end{aligned} \tag{3.53}$$

where we calculate the difference between the gravitational attraction at the center and the closer edge of one of the objects. If we can assume that $\Delta r \lll r$, we need retain only first order terms; dividing by r^2 we then get

$$\begin{aligned} \Delta F &\approx -G\mathfrak{M}m[2\Delta r/r][r^2 - 2r\Delta r]^{-1} \\ &\approx 2G\mathfrak{M}m\Delta r/r^3 \end{aligned} \tag{3.54}$$

The result can be obtained more elegantly by use of the *del* or *nabla* operator as

$$\Delta F = \nabla(-G\mathfrak{M}mr^{-2}) = 2G\mathfrak{M}m\ \Delta r/r^3 \tag{3.55}$$

If we are examining the bulge raised in the Earth by a satellite, $\Delta r = R_\oplus$. Because the distance of the Moon varies by $\pm ae = \pm 0.055a$, the relative change in the tide-raising force is

$$\delta(\Delta F)/\Delta F \approx 0.165$$

Example 3.2 Relative tides due to the Sun and Moon on the Earth

The tidal effect on the Earth by the Moon is currently about twice that by the Sun:

$$\Delta F_{\oplus-\odot} = -2Gm_\oplus\mathfrak{M}_\odot\ R_\oplus/{a_\oplus}^3 \tag{3.56}$$

while

$$\Delta F_{\oplus-☽} = -2Gm_\oplus m_☽\ R_\oplus/{a_☽}^3 \tag{3.57}$$

so that

$$\begin{aligned} \Delta F_{\oplus-☽}/\Delta F_{\oplus-\odot} &= [m_☽/\mathfrak{M}_\odot](a_\oplus/a_☽)^3 \\ &= [7.35 \times 10^{22}/2.0 \times 10^{30}](1.5 \times 10^{11}/3.8 \times 10^8)^3 = 2.2 \end{aligned}$$

Notice that there are bulges raised on both sides of the planet by tidal action. The far side is accelerated less than, say, the planetary core, while the core is accelerated less than the near side.

Because of friction, however, the tidal bulges may not be able to localize to the sub-satellite position on the planet. If the angular speed of rotation exceeds the angular motion of the satellite in its orbit, and is in the same direction, then friction may result in the tide being swept along with the rotation, causing the bulge to precede the satellite. This is the case for the Earth (see Figure 3.6, where the tides have been greatly exaggerated for visibility). The net effect is for the satellite to drag on the near-side bulge and to accelerate the far-side bulge. The slightly greater proximity of the near-side bulge (by the inverse square law) results in the torque on the near side being larger than that on the far side. The differential force again goes as r^{-3}. This causes a net braking of the rotation of the planet. The evidence for the slowing of the Earth's rotation is overwhelming, as we note in the next section.

3.7.2 Effect of Tidal Friction on Orbits

The same bulge that gets braked in the discussion in the preceding section also accelerates the Moon. The acceleration increases the instantaneous orbital speed, resulting in a slight increase in the semi-major axis. As the orbit enlarges the orbital speed decreases. Thus the second consequence of tidal friction is to increase the orbit of the Moon, and thus, by Kepler's 3rd law, its period.

The evidence for secular variation in the semi-major axis of the Moon is also strong. Consolmagno and Schaefer (1994, p. 246) cite evidence that in the Devonian Period [~400 million years (My) before present] there were only 10 days in the lunar month, with an average change in the length of the day since then of 25^{s}/My. They estimate that the Moon at that time would have been half as distant as it is now. The rate of recession depends strongly on the effectiveness of tidal friction, which in the current epoch is concentrated in shallow sea passages such as the Bering Strait and the English Channel. This means that the rate must vary over time: shallow seas come and go over geological time due to continental drift.

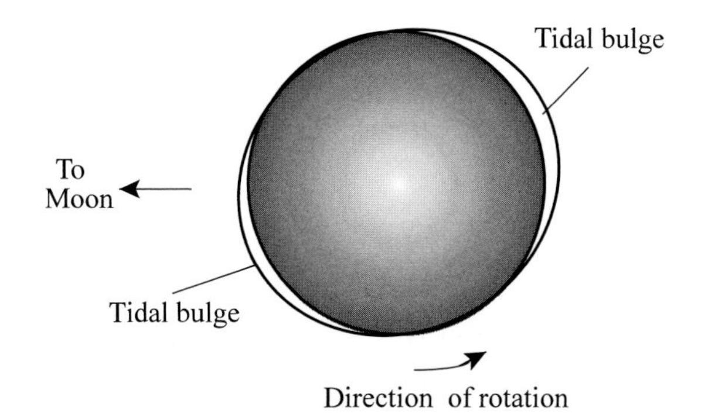

Fig. 3.6. Tidal braking of the Earth by lunar torques

Therefore, continental drift has changed the rate of spindown of the Earth and the rate of recession of the Moon. Coupled to this must be the mean ocean level, which, in turn, depends on the state of glaciation and ice cap thickness.

The end of the current trend (Earth rotation slowing; Moon receding) will occur when the Earth and Moon become tidally locked, with the Earth day and lunar month being of approximately equal length. However, this equilibrium situation may not last. The tidal action of the Sun is to slow the Earth's rotation also. Once the Earth–Moon lock becomes broken, the slower Earth may brake the Moon's motion, resulting in an inward spiraling Moon. Such a decay can be seen in Phobos' orbit about Mars.

3.7.3 Resonances and Commensurabilities

Tide-raising effectiveness must change with time as the distance increases. When the rotation and revolution are in some type of synchronism, a spin–orbit resonance is said to occur. An example is seen in the 1:1 ratio of revolution to rotation of Earth's Moon. However, the ratio need not be 1. For Mercury, the ratio is 2:3.

There are also orbit–orbit resonances. The classical example of the Galilean satellites was discovered by Pierre–Simon de Laplace (1749–1827) in 1805. In this case,

$P_{\text{Ganymede}} = 2P_{\text{Europa}}$, and $P_{\text{Europa}} = 2P_{\text{Io}}$. Expressing the mean motion as $n = 2\pi/P$, the mutual locking can be described by the relation

$$n_{\text{Io}} - 3n_{\text{Europa}} + 2n_{\text{Ganymede}} = 0 \tag{3.58}$$

There is a host of resonances among Saturn's moons. Tables 3.1 and 3.2 list the best known spin–orbit and orbit–orbit coupling cases in the solar system.

Among longer period effects is the *long-period inequality* between Jupiter and Saturn; their motions are not quite commensurable, but very close:

$$5n_{\text{Saturn}} - 2n_{\text{Jupiter}} = 3''.99/d \tag{3.59}$$

Table 3.1. Some solar system spin–orbit resonances

Objects	e	i	P_{rtn}	P_{rev} (Sid.)	P_{rev} (Syn)	$P_1/P_2 = ?$
☿-⊕	0.206	7°.003	59d	87^d^.969	115^d^.88	0.509 ≈ 1/2(rtn/syn)
☿-☉						0.671 ≈ 2/3(rtn/sid)
♀-⊕	0.007	3°.395	244.3	224.701	583.92	0.418 Earth ≈ 5/12(rtn/syn)

Table 3.2. Some solar system orbit–orbit resonances

Objects	e	i	P_{rev}	$P_1/P_2 =?$
Earth	0.017	$0^\circ\!.0$	1.0y	$0.625 \approx 5/8$
Toro	0.435	$9^\circ\!.3$	1.6	
Earth	0.017	$0^\circ\!.0$	1.0	$0.393 \approx 2/5$
Ivar	0.397	$8^\circ\!.3$	2.545	
Jupiter	0.048	$1^\circ\!.4$	11.86	$1.501 \approx 3/2$
Hilda	0.15	$7^\circ\!.9$	7.90	
Jupiter	0.048	$1^\circ\!.4$	11.86	$1.333 \approx 4/3$
Thule	0.03	$23^\circ\!.$	8.90	
Jupiter	0.048	$1^\circ\!.4$	11.86	1
Trojans	0.15:	15:	11.86	
Tethys	0.00	$1^\circ\!.1$	$1^{\text{d}}\!.887802$	2.003
Mimas	0.020	$1^\circ\!.5$	0.942422	
Dione	0.002	$0^\circ\!.0$	$2^{\text{d}}\!.93681$	1.997
Enceladus	0.0045	$0^\circ\!.0$	1.37028	
Hyperion	0.104	$0^\circ\!.5$	$21^{\text{d}}\!.27666$	$1.334 \approx 4/3$
Titan	0.0290	$0^\circ\!.3$	15.945452	
Pluto	0.247	$17^\circ\!.1$	248.43	$1.508 \approx 3/2$
Neptune	0.0087	$1^\circ\!.5$	164.78	

where $n_{\text{Jupiter}} = 299''\!.13/\text{d}$ and $n_{\text{Saturn}} = 120''\!.45/d$. Because this inequality involves mean motions, and in one revolution, $n = 1,296,000''/P(\text{y})$, one may also write

$$5P_{\text{J}} - 2P_{\text{S}} \triangleq \kappa P_{\text{J}} P_{\text{S}} \tag{3.60}$$

where $P_{\text{J}} = 11^{\text{y}}\!.862$ and $P_{\text{S}} = 29^{\text{y}}\!.459$, and $\kappa = 0.00112$. The terms on the LHS have the values $58^{\text{y}}\!.918$ and $59^{\text{y}}\!.310$, respectively. Thus, Jupiter and Saturn will come to the same heliocentric configuration about every 59 years.

In the Saturn system there are a number of commensurabilities. For instance, $n_{\text{titan}} = 1.334342\ n_{\text{Hyperion}}$, very close to a 4:3 resonance.

In some cases the perturbations on small objects are sufficient to lock them in or sometimes out of certain orbits. Dione and Tethys have a 1:1 resonance with small objects in their respective orbits, and Janus and Epimetheus have horseshoe orbits and interchange orbits, and are also locked into a 1:1 orbital resonance. The moons of Saturn sweep away material from certain ring tori, leaving gaps. In the Uranian system, Rosalind and Cordelia have a 5:3 resonance. Cordelia and Ophelia provide bounds to the $\in$ ring, with 24:25 and 14:13 resonances. Pluto and Charon are also locked into a 1:1 resonance.

Jupiter perturbs the orbits of minor planets with sub-multiples of Jupiter's period. The "Kirkwood gaps" in the asteroid belt are the result.

3.8 Perturbations

3.8.1 Causes and Effects

Sources of perturbations include other bodies, non-spherical distributions of the mass of the objects in the two-body system or both, and viscous media or other sources of drag. The orbital elements of an object in a two-body orbit subject to a point source mass will remain constant. However, as noted in Section 3.2, subject to perturbations they will undergo variation. The variation of each element depends on the relative direction and magnitude of the perturbing force. At some instant, the object in such a perturbed orbit may be said to have an *osculating orbit* because it "kisses" the true path at that instant (and perhaps at none other). If all perturbing forces were to disappear at that instant (and remained absent thereafter), the orbital motion thenceforth would be described accurately by the elements of the osculating orbit.

Relative to the two-body orbital plane, perturbations that are said to be

1. *Normal*, alter the inclination (i) (and, through precession, Ω and ω);
2. *Radial*, alter the eccentricity (e), and possibly the semi-major axis (a); and
3. *Transverse* to the other two directions, alter a and e.

Transitions to/from transfer orbits, when a thrust is given the spacecraft at either pericentre or apocentre, involve perturbing effect 3.

Satellite orbits undergoing perturbations are subject to effects 1 and 2, and precess.

We now determine the changes in the osculating elements due to a general perturbation force, but which we assume to be small compared to the principal two-body force. This is usually not an unrealistic assumption, given the relatively large separations of bodies and the dominance of the Sun's gravitational effects in the solar system. An excellent detailed source for this treatment is Murray and Dermott (1999/2001), which our summary follows.

3.8.2 Deriving the Variation in the Orbital Elements

A small disturbing force per unit mass, dF, sometimes referred to as the *perturbative acceleration*, may be expressed in terms of its components:

$$\mathrm{d}\mathbf{F} = R\hat{\mathbf{r}} + T\,\hat{\boldsymbol{\theta}} + N\,\hat{\mathbf{z}} \tag{3.61}$$

where $\hat{\mathbf{r}}$ and $\hat{\theta}$ are unit vectors as defined in Section 3.1 and $\hat{\mathbf{z}}$ is in the direction normal to the orbit; and R, T, and N are the components of the perturbative force in the radial, transverse, and normal directions, respectively.

In this case, the energy of the orbit is no longer constant. The time-variation is

$$\dot{E} = \dot{\mathbf{r}} \cdot \mathrm{d}\mathbf{F} = \dot{r}R + r\,\dot{\theta}\,T \tag{3.62}$$

As per equation (3.36), $E = -(1/2)\mu/a$, so the variation in a is

$$\dot{a} = 2a^2\,\dot{E}/\mu \tag{3.63}$$

Rewriting equation (3.25) for the ellipse explicitly, we get

$$r = a(1-e^2)/[e\,\cos\,\theta + 1] \tag{3.64}$$

and taking the time derivative due to motion in the orbit (a and e kept constant),

$$\dot{\mathrm{r}} = r\,\dot{\theta}\,e\,\sin\,\theta/[1 + e\cos\theta] \tag{3.65}$$

Setting $\dot{\theta} = \dot{\upsilon}$ (i.e., noting that the orbit angle variation is that of the true anomaly),

$$\dot{\mathrm{r}} = r\,\dot{\upsilon}\,e\sin\upsilon/[1 + e\cos\upsilon] \tag{3.66}$$

and with $(r\,\dot{\upsilon})^2 = h^2/r^2$,
and $h^2 = a\,\mu(1-e^2) = n^2a^4\,(1-e^2)$,
where $n = 2\pi/P = [\mu/a^3]^{1/2}$ [from (3.39), (3.40)], we can show that

$$\dot{\mathrm{r}} = n\,a\,e\sin\upsilon/(1-e^2)^{1/2} \tag{3.67}$$

and

$$r\,\dot{\upsilon} = n\,a\,[1 + e\cos\upsilon]/(1-e^2)^{1/2} \tag{3.68}$$

Substituting equations (3.67) and (3.68) into (3.62), from (3.63), we derive

$$\dot{a} = \{2/[n\sqrt{(1-e^2)}]\}[(e\sin\upsilon)\,R + (1 + e\cos\upsilon)\,T]$$

or, with a explicit,

$$\longrightarrow \qquad \dot{a} = \{2\,a^{3/2}/[\sqrt{\mu}(1-e^2)]\}[(e\sin v)\,R + (1+e\cos v)\,T] \qquad (3.69)$$

From the relationship between h and $\sqrt{a}(1-e^2)$, and equation (3.40), we get

$$e = \{1 - h^2/(\mu a)\}^{1/2} \qquad (3.70)$$

whence,

$$\dot{e} = 1/2\ \{1 - h^2/(\mu a)\}^{-1/2}[-2h\dot{h}/(\mu a) + h^2\,\dot{a}/(\mu a^2)] \qquad (3.71)$$

We know $\dot{a}$ from (3.69) but we must find $\dot{\mathbf{h}}$. Because

$$\mathbf{h} = \mathbf{r} \times \dot{\mathbf{r}}$$

it follows that

$$\dot{\mathbf{h}} = \dot{\mathbf{r}} \times \dot{\mathbf{r}} + \mathbf{r} \times \ddot{\mathbf{r}} = 0 + \mathbf{r} \times \mathrm{d}\mathbf{F} \qquad (3.72)$$

and because

$$\mathbf{r} = r\hat{\mathbf{r}} \qquad (3.73)$$

with equation (3.61),

$$\mathbf{h} = r\hat{\mathbf{r}} \times (R\,\hat{\mathbf{r}} + T\,\hat{\boldsymbol{\theta}} + N\,\hat{\mathbf{z}}) \qquad (3.74)$$

so that

$$\dot{\mathbf{h}} = r\,T\,\hat{\mathbf{z}} - r\,N\,\hat{\boldsymbol{\theta}} \qquad (3.75)$$

However, the $r\,N\,\hat{\boldsymbol{\theta}}$ term alters the direction but not the magnitude of $\mathbf{h}$, hence

$$\dot{h} = r\,T \qquad (3.76)$$

Substituting (3.76), the lower equation of (3.69), and the relation, $h^2 = a\,\mu(1-e^2)$, from (3.24), into (3.71), after some simplification, we arrive at

$$\longrightarrow \qquad \dot{e} = [a(1-e^2)/\mu]^{1/2}\ [R\sin v + T\,(\cos E + \cos v)] \qquad (3.77)$$

where E is the eccentric anomaly. Note that the variation of the eccentricity depends only on the components of the perturbation in the orbital plane.

The variation of the inclination, i, is more complicated. We begin with the components of $\mathbf{h}$ in the orthogonal directions $\hat{\mathbf{x}}$, $\hat{\mathbf{y}}$, $\hat{\mathbf{z}}$ in which $\hat{\mathbf{x}}$ is in the direction of the vernal equinox, $\hat{\mathbf{y}}$ is 90° in the direction of increasing longitudinal coordinate, in the reference plane, and $\hat{\mathbf{z}}$ is normal to the plane. With $h_z = \mathbf{h} \cdot \hat{\mathbf{z}}$, etc.,

$$\begin{aligned} h_x &= h \sin i \sin \Omega \\ h_y &= -h \sin i \cos \Omega \\ h_z &= h \cos i \end{aligned} \tag{3.78}$$

Taking the derivative of h_z,

$$\dot{h}_z = \dot{h} \cos i - h \sin i \; i' \tag{3.79}$$

where $i' \equiv \mathrm{d}i/\mathrm{d}t$, so

$$i' = [\dot{h} \cos i - \dot{h}_z]/(h \sin i)$$

Simplifying,

$$\begin{aligned} i' &= [(\dot{h}/h) \cos i - \dot{h}_z \cos i/h_z]/ \sin i \\ i' &= [\dot{h}/h - \dot{h}_z/h_z]/\sqrt{[(1 - \cos^2 i)/ \cos^2 i]} \end{aligned}$$

From (3.78), we get

$$i' = [\dot{h}/h - \dot{h}_z/h_z]/\sqrt{[(h/h_z)^2 - 1]} \tag{3.80}$$

The derivatives of h_x, h_y, and h_z may be shown to be

$$\begin{aligned} \dot{h}_x &= r[T \sin i \sin \Omega + N \sin(\omega + v) \cos \Omega + N \cos(\omega + v) \cos i \sin \Omega] \\ \dot{h}_y &= r[-T \sin i \cos \Omega + N \sin(\omega + v) \sin \Omega - N \cos(\omega + v) \cos i \cos \Omega] \\ \dot{h}_z &= r[T \cos i - N \cos(\omega + v) \sin i] \end{aligned} \tag{3.81}$$

which involves transformations from the orbital plane into the reference plane. Note that the argument $(\omega + v)$ is the argument of the latitude, u, mentioned in Section 3.4, and it is measured from the node. Ultimately, we get

$$i' = r\, N \cos(\omega + v)/h$$

or

$$\longrightarrow \qquad i' = \{[a(1 - e^2)/\mu]^{1/2}\, N \cos(\omega + v)\}/[1 + e \cos v] \tag{3.82}$$

Note that only the component of the perturbing force normal to the orbital plane has an effect on i.

The effect on the longitude of the ascending node is obtained by dividing the first by the second equation of (3.78)

$$h_x/h_y = -\tan\,\Omega \tag{3.83}$$

then, taking the derivative,

$$-\sec^2\,\Omega\cdot\dot{\Omega} = \dot{h}_x/h_y - h_x\,\dot{h}_y/{h_y}^2$$

after manipulation,

$$\dot{\Omega} = (h_x\dot{h}_y - h_y\dot{h}_x)/(h^2 - {h_z}^2) \tag{3.84}$$

and, after appropriate substitutions,

$$\dot{\Omega} = r\,N\,\sin\,(\omega + \upsilon)/(h\,\sin\,i)$$

or

$$\longrightarrow \qquad \dot{\Omega} = \{[a(1-e^2)/\mu]^{1/2}\,N\,\sin\,(\omega+\upsilon)\}/[\sin\,i\,(1+e\,\cos\,\upsilon)] \tag{3.85}$$

Note the sensitivity of $\dot{\Omega}$ only to the N component of d**F**. This component causes the motion of the plane of the orbit.

We can find the variation of ω by the substitution $\upsilon = u - \omega$ in the equation of the ellipse, where u is the argument of the latitude, referred to above and in Section 3.4. The expression takes on the form

$$h^2 = \mu r\{1 + \surd[1 + 2\,E\,h^2/\mu^2]\cos\,(u-\omega)\} \tag{3.86}$$

Now we need to take the time derivative. For this exercise, we are interested in the instantaneous value of $\dot{\omega}$ due to the perturbing force, so r is treated as a constant even though E, h, and ω are allowed to vary. The result is

$$\begin{aligned}\dot{\omega} = {} & 2h\,\dot{h}[(1/r) + E/(e\mu)\cos\,(u-\omega)]/[e\mu\,\sin\,(u-\omega)] + \dot{u} \\ & - [h/(e\mu)]^2\,\dot{E}\,\cot\,(u-\omega)\end{aligned} \tag{3.87}$$

Making use of substitutions for $\dot{h}$, $\dot{E}$ [and then $\dot{a}$ from (3.63)] as given above, we get

$$\longrightarrow \qquad \begin{aligned}\dot{\omega} = {} & (1/e)[a\,(1-e^2)/\mu]^{1/2}\{-R\,\cos\,\upsilon \\ & + T\,\sin\,\upsilon\,[(2+e\,\cos\,\upsilon)/(1+e\,\cos\,\upsilon)]\} - \dot{\Omega}\,\cos\,i\end{aligned} \tag{3.88}$$

The last term, involving $\dot{\Omega}$, arises from the changing position of the nodes, from which u is measured.

The effect of a perturbation on the epoch (the instant that the object is at a particular point in the orbit, such as the pericenter), can be determined by differentiation of Kepler's equation, defined in (3.46), and rewritten here as

$$M = \varepsilon - e \sin \varepsilon \tag{3.89}$$

where, ε is the eccentric anomaly as we defined it in (3.46), but to avoid confusion with the energy per unit mass in this section, we make the substitution $\epsilon = E$ for the eccentric anomaly. Now $M = n(t - t_0)$, and taking the derivative, we obtain

$$\dot{M} = \dot{n}t + n - \dot{n}t_0 - n\dot{t}_0 = \dot{\epsilon} - \dot{e} \sin \epsilon - e\,\dot{\epsilon} \cos \epsilon \tag{3.90}$$

Thus,

$$\dot{t}_0 = (1/n)\{ - \dot{n}t_0 + n + \dot{n}t - \dot{\epsilon}\,(1 - e \cos \epsilon) + \dot{e} \sin \epsilon$$

or

$$\dot{t}_0 = (1/n)[\dot{n}(t - t_0) + n - \dot{\epsilon}\,(1 - e \cos \epsilon) + \dot{e} \sin \epsilon] \tag{3.91}$$

Differentiating (3.40),

$$\begin{aligned} \dot{n} &= D(\mu/a^3)^{1/2} = 1/2(\mu/a^3)^{-1/2}(-3\mu\, a^{-4})\dot{a} \\ &= -(3/2)(\mu/a^3)(\dot{a}/a)/(\mu/a^3)^{1/2} = -(3/2)n\,\dot{a}/a \end{aligned} \tag{3.92}$$

and $\dot{a}$ is known from (3.69), so

$$\begin{aligned} \dot{n} = &-(3/2)(n/a)\{2\,a^{3/2}/[\sqrt{\mu}(1 - e^2)]\} \\ &\times [(e \sin v)R + (1 + e \cos v)T] \end{aligned} \tag{3.93}$$

The quantity $\dot{\epsilon}$ in (3.90) and (3.91) is found from equation (3.47), where the eccentric anomaly was represented by the symbol E. Rewriting it,

$$\dot{\epsilon} = n/[1 - e \cos \epsilon]$$

Thus the second and third terms of (3.91) vanish. Finally, $\dot{e}$ is known from (3.77). Substitution into (3.91), and rearranging, results in the following:

$$\longrightarrow \quad \begin{aligned} \dot{t}_0 = &\Big\{\{ - 3(t - t_0)/[\mu(1 - e^2)/a]^{1/2}\}e \sin v \\ &+ [a^2(1 - e^2)^{1/2}/\mu] \sin v\Big\}R \\ &+ \{\{ - 3(t - t_0)/[\mu(1 - e^2)/a]^{1/2}\}[1 + e \cos v] \\ &+ [a^2(1 - e^2)^{1/2}/\mu][\cos v + \cos \epsilon]\}\,T \end{aligned} \tag{3.94}$$

Note that only the components of the perturbing force in the plane of the orbit have a direct effect on the epoch. Note also that the effect grows with time because of the $(t - t_0)$ term on the RHS of (3.94). A more sophisticated treatment of the effect on the epoch can be found in Brouwer and Clemence (1961, pp. 285–289), Danby (1962, p. 243), and Murray and Dermott (1999/2001, p. 57).

For our purposes, it suffices that we have demonstrated how perturbations can affect the elements of the orbit. It is beyond the scope of the present work to discuss how to find the perturbing force given the change to the elements in general, but we will discuss later the determination of the distribution of mass in a planet given the perturbations of the orbit of a satellite. Other topics in, and techniques of, celestial mechanics can be obtained from the three excellent sources mentioned in the previous paragraph. As a practical interest, many programs (written in BASIC) are provided in Danby (1988).

This concludes our brief summary of celestial mechanics. We now turn to the physical components of the solar system, starting with its dominant member, the Sun.

References

Applegate, J. H., Douglas, M. R., Gursel, Y., Hunter, P., Seitz, C. L., and Sussman, G. J. 1985. "A Special Purpose Multiprocessor to Compute Planetary Orbits," *IEEE Transactions on Computers*, **C**-34, No. 9, 822–831.

Bradskect, D. H. and Steelman, D. P. 2004. *Binary Maker 3.0,* Contact Software, Norristown, PA 19087.

Brouwer, D. and Clemence, G. M. 1961. *Methods of Celestial Mechanics* (New York: Academic Press).

Cohen, E. R. and Giacomo, P. 1987. *Symbols, Units, Nomenclature and Fundamental Constants in Physics* (Int. Union of Pure & Applied Physics) (Reprint of *Physica*, **146**A, (1987) pp. 1–68).

Consolmagno, G. J. and Schaefer, M. W. 1994. *Worlds Apart: A Textbook in the Planetary Sciences* (Englewood Cliffs, NJ: Prentice Hall).

Danby, J. M. A. 1962. *Fundamentals of Celestial Mechanics*, first Ed. (New York: McMillan) (Reprinted in 1964).

Danby, J. M. A. 1988. *Fundamentals of Celestial Mechanics*, second Ed. (Richmond, VA: Willmann-Bell).

Milone, E. F., and Wilson, W. J. F. 2008. *Solar System Astrophysics: Planetary Atmospheres and the Outer Solar System* (New York: Springer)

Montenbruck, O. 1989. *Practical Ephemeris Calculations* (New York: Springer) (Tr. by A. H. Armstrong of Montenbruck's *Grundlagen der Ephemeridenrechnung*, 3, (Munich: Verlag Sterne und Weltraum Dr Vehrenberg GmbH), 1987).

Moulton, F. R. 1914. *An Introduction to Celestial Mechanics*, second Revised Ed. (New York: MacMillan).

Murray, C. D. and Dermott, S. F. 1999/2001. *Solar System Dynamics* (Cambridge: University Press).
Neutsch, W. and Scherer, K. 1992. *Celestial Mechanics: An Introduction to Classical and Contemporary Methods* (Mannheim: B.I. Wissenschaftsverlag).
Peirce, B. O. 1957. *A Short Table of Integrals*, 4th Ed. (Boston, MA: Ginn and Co.).
Schlosser, W., Schmidt-Kaler, Th., and Milone, E. F. 1991/1994. *Challenges of Astronomy: Hands-On Experiments for the Sky and Laboratory* (New York: Springer-Verlag).
Smart, W. H. 1953. *Celestial Mechanics* (New York: Wiley).
Szebehely, V. G. 1989. *Adventures in Celestial Mechanics* (Austin, TX: University of Texas Press).

Challenges

[3.1] Derive Kepler's third law from Newton's gravitational and motion laws. [Hint: Consider areal speed and make use of equation (2.37).]

[3.2] Show that $\mu = G(M + m)$ for relative orbits in a two-body system.

[3.3] Derive the relationship between r and θ for an attractive inverse square force law. (Solution is given in the Appendix).

[3.4] Assume that the initial orbit for an Earth satellite is 100 km above the Earth's (mean) equator and that its final orbital radius is that of a geo-synchronous satellite. (a) Compute the orbital elements and other parameters for the Earth satellite transfer orbit. Compute the velocities of (b) the circular orbits and (c) the transfer orbit at points of thrust and thus the velocity difference required to achieve the changed orbit. For these purposes, you can ignore other perturbations.

[3.5] (a) Compare the orbital properties of synchronous satellites on Mars and Earth and (b) compute the orbit of a synchronous satellite of the Moon. (c) Demonstrate the feasibility or non-feasibility of such a lunar satellite.

[3.6] There have been discussions about spacecraft paths that make full use of the gravitational attraction of solar system objects and thus minimize thrusts and use of chemical fuel. (a) Discuss the celestial mechanics involved in the design of such a 'highway' to the outer planets; (b) write down an expression for the net acceleration on the spacecraft at some instant; and (c) describe an iterative process which can be used to predict its future path without additional use of its rockets.

[3.7] If we now know the masses of all the planets to high precision, why is it difficult to predict the exact positions of Earth-crossing asteroids a few decades into the future?

4. The Core of the Solar System: The Sun

The planets and other features of the solar system are, as this name implies, dominated by the Sun. Therefore, we begin by placing this primary component in context, describe its properties as a star, and discuss the relevant astrophysics required to gain some insight into its nature and importance.

4.1 The Solar Context

The role of the Sun in our solar system can be demonstrated in these empirical data:

- Visually the Sun is the solar system's dominant object; it is a star, a self-luminous body, powered by nuclear reactions in its core.
- The Sun is the most massive object in the solar system.
- Nevertheless, the angular momentum is concentrated in the planets, mainly Jupiter.
- The planets all revolve in the same direction (CCW as viewed from above the north ecliptic pole), the same direction as the rotation of the Sun.
- The orbits are roughly coplanar, except for those of Mercury and groups of small-body and "dwarf planet" objects, such as Pluto, in the outer solar system.
- The orbits are roughly circular, except for those of Mercury and groups of small-body and "dwarf planet" objects, such as Pluto, in the outer solar system.
- There is a debris field (asteroid belt) and evidence for clouds of (cometary) debris beyond the orbit of Neptune.
- There is evidence for differentiation in chemical composition across the solar system.
- The Titius–Bode "law" suggests an underlying—although perhaps incompletely realized—principle or scheme for the spacing of the planets' (and

the largest asteroid's) orbits, as noted in Chapter 1. See Nieto (1972) for an extensive discussion.

These points are discussed in other contexts in other chapters; for now it suffices to state that all theories of the origin of the solar system must take these observations into account. In our investigation, the properties of each group of solar system objects will be examined, and in the end we will try to see how our perception of the solar system fits into a more general context being formulated after the discoveries of planets around other stars. We begin at our local center, with the Sun.

The Sun was revered as a god for its life-sustaining warmth and its light long before its gravitational dominance was recognized. The apparent motion of the Sun along the ecliptic and the resulting seasonal variation in insolation and thus warmth was recognized at least 6 millennia ago, and probably much earlier, as noted in Chapter 1. The phenomenological effects of the Earth's revolution and rotation can be probed with the tools described in Chapters 2 and 3. We discuss in this chapter the radiative properties of sunlight and how light interacts with matter. The explanation of the observable properties of the Sun requires a brief summary of radiation laws and the review of a number of definitions. The black body radiation laws will be described amid the descriptive properties of the Sun and other stars. Neither the Sun nor any other star is a perfect black body radiator, but it is often convenient to compare their properties to those of black bodies and to try to understand the causes of the differences. The light and particle emissions of the Sun are critical to the understanding of planetary properties and phenomena, and so these emissions and their variations must be described also.

4.2 The Sun as a Star in the Milky Way Galaxy

The average distance of the Sun from the Earth is the value of the *astronomical unit*: $1.495979(1) \times 10^{11}$ m (the number in parentheses is the uncertainty in units of the last decimal place). This is sometimes expressed as the solar parallax, the *mean equatorial horizontal parallax*,

$$8''.79418 = 4.26354 \times 10^{-5} \text{ radian},$$

the displacement shift the Sun would appear to undergo at mean distance, as viewed alternately from the center and from the limb (horizon) of Earth, using the equatorial radius as the baseline. These and most of the other data in this section are taken from Allen (1973, p. 161ff), Cox (2000, p. 340ff), and the 2005 Astronomical Almanac, with occasional updates from other sources.

The mass of the Sun can be determined from a formulation of Kepler's third law,[1] the length of the year, and the mean distance of the Earth from the Sun,

$$\mathfrak{M}_\odot = 1.989(2) \times 10^{30}\,\text{kg}$$

This mass represents 99.9% of the total mass of the solar system. From the mass and the radius of the Sun, we find its mean density,

$$\begin{aligned} < \rho_\odot > &= \mathfrak{M}_\odot/[4/3\,\pi\mathfrak{R}_\odot{}^3] \\ &= 1409\,\text{kg/m}^3 = 1.409\,\text{g/cm}^3 \end{aligned} \tag{4.1}$$

and the gravitational acceleration at the solar radius $\mathfrak{R}_\odot$,

$$g_\odot = G\mathfrak{M}_\odot/\mathfrak{R}_\odot{}^2 \tag{4.2}$$

so that $g_\odot = 273.98(4)\,\text{m/s}^2 = 2.7398(4) \times 10^4\,\text{cm/s}^2$

Another important physical property is the solar *angular momentum*,

$$\begin{aligned} \mathbf{L}_\odot &= \sum \mathfrak{M} \cdot \mathbf{v} \times \mathbf{r} = \mathbf{I} \cdot \omega \\ &= 1.63 \times 10^{48}\,\text{g cm}^2/\text{s} \end{aligned} \tag{4.3}$$

Here $\mathfrak{M}$, $\mathbf{v}$, and $\mathbf{r}$ are the mass, velocity due to rotation, and the distance from the rotation axis, respectively, and the summation is taken over all solar atoms and ions! I is the *moment of inertia*, which for a uniform sphere of mass $\mathtt{M}$ and radius $\mathfrak{R}$ is

$$I_{\text{uniform sphere}} = 2/5 \cdot \mathtt{M}\mathfrak{R}^2 \tag{4.4}$$

and where

$$\omega = 2\pi/P_{\text{rotn}} \tag{4.5}$$

is the mean *angular velocity* and P_{rotn} is the rotation period. The value for $\mathbf{L}_\odot$ is based on the reasonable assumption that the internal rotation is the same as at the surface, as suggested by helioseismology data (Beatty and Chaikin 1990, p. 24). The overall angular momentum of the entire solar system is, however,

$$L_{\text{total}} = 3.148 \times 10^{50}\,\text{g cm}^2/\text{s}, \text{ so that } \mathrm{L}_\odot/L_{\text{total}} \approx 0.005$$

Thus the angular momentum of the Sun makes up a negligible part of the total angular momentum.

---

[1] $\mathrm{G}(\mathfrak{M}_\odot + \mathfrak{M}_\oplus)\,\mathrm{P}^2 = 4\pi^2 a^3$, where the gravitational constant, $G = 6.6726(9) \times 10^{-11}\,\text{m}^3/\text{kg s}^2$, $\mathfrak{M}_\odot$ and $\mathfrak{M}_\oplus$ are the masses of the Sun and Earth, the orbital period of the Earth, $P = 3.1558 \times 10^7$ s, and the semi-major axis of Earth's orbit, $a = 1.4960 \times 10^{11}$ m.

The angular semi-diameter of the Sun in a narrow passband of red light ($\lambda = 0.800\,\mu$m) was determined from a 6-year series of daily ground-based measurements to be $\alpha = 959''.680(9) = 0.00465266(4)$ radian (Brown and Christensen-Dalsgaard 1998). The apparent solar radius is found, after adjusting the distance for the position of the observatory relative to the center of the Earth, and for the difference between the barycenter of the Earth–Sun system and the center of the Sun, to be

$$\begin{aligned}\Re_\odot &= r\alpha = 1.495936 \times 10^{11}\,\text{m} \times 4.65266(4) \times 10^{-3} \\ &= 6.96008(7) \times 10^{8}\,\text{m}\end{aligned}$$

This is slightly larger than the previously accepted value (6.9599×10^8 m) for the solar radius. However, for many purposes the important radius is not the apparent edge of the Sun at a particular wavelength, but the radius at which the local temperature of the Sun is equal to the effective temperature (defined in Section 4.5). The correction requires a solar model. From the mean of the radii corrected with two such models, Brown and Christensen-Dalsgaard (1998) obtained for this definition of the solar radius,

$$\Re_\odot = 6.95508(26) \times 10^{8}\,\text{m}$$

The departure of the Sun's geometric figure from a sphere is small, but measureable. The *oblateness*, ϵ, is the difference between equatorial and polar radius in units of the equatorial radius. From highly precise measurements obtained in 1992 and 1994 with a balloon-borne instrument (Lydon and Sofia, 1996), a weighted mean value of $\epsilon = 8.92(78) \times 10^{-6}$ is obtained.

The stars in the vicinity of the Sun, kinematically defining the *local standard of rest* (LSR), are moving through space in a circular orbit at a speed of ~220 km/s and are located about 8.5 Kpc from the center of the galaxy. These values were formally adopted by the IAU in 1985. The "solar motion" is the velocity of the Sun with respect to the LSR. The determined components of this motion in the directions away from the galactic center (u), in the direction of galactic rotation (v), and normal to the galactic plane positive toward the North Galactic Pole (NGP) (w) are:

$$\begin{aligned}u_\odot &= -9\,\text{km/s},\\ v_\odot &= +12\,\text{km/s}\\ w_\odot &= +7\,\text{km/s}\end{aligned}$$

The net motion is 16.5 km/s, directed toward the position: $\ell = 53°$, $b = 25°$ in Hercules, a direction known as the solar apex. Note that the components indicate that the Sun is approaching the galactic center, is slightly faster at present than the appropriate orbital speed for a circular orbit at the current distance of the Sun, and also has motion normal to the plane. Thus the Sun has a non-circular and non-planar orbit about the galactic center. It is at the edge of a spiral arm and its basic properties identify it as a member of the "old disk" population. Its age is estimated at 5×10^9 years.

4.3 Observable Properties of the Quiet Sun

The title of this section reveals that the Sun is a kind of Dr Jekyll and Mr Hyde. In this section we discuss its gentler nature, but the Mr Hyde aspect (the "active" Sun) is never more than half a solar (or "sunspot") cycle away. Direct imaging observations across the electromagnetic spectrum, in addition to eclipse observations, reveal that the Sun's outer regions consist of three parts:

The *photosphere* (literally "sphere of light")
The *chromosphere* ("sphere of color")
The *corona* or halo

The chromosphere and corona are best seen during total solar eclipses, either natural or artificially produced eclipses, with instruments known as *coronographs*, but the eclipse condition is not strictly necessary. Direct images of the Sun, taken with a type of spectrograph known as a *spectroheliograph*, in very narrow passbands, can be recorded as *spectroheliograms*, pictures of the Sun at higher regions of the atmosphere than can be seen in "white light" (very broad passband) images.

The motions of neutral and ionized atoms in the Sun can be studied through the analysis of the profiles of spectral lines. Contributions to the width of a spectral line come from several sources: (a) intrinsically, from the uncertainty principle; (b) from the relative abundance of the atoms in the particular ionization stage and excitation states involved in the atomic transition; and (c) the motion and the pressure in the atmosphere. Some of this motion arises from the kinetic energy of the atoms and is therefore dependent on the temperature. Other motions are due to the turbulence on small spatial scales (*microturbulence*) and on larger scales (*macroturbulence*) or the bulk motion of large areas of the Sun (*granulation*, *supergranulation*, and rotation). Spectrographs at higher spectral and spatial resolution are used to sort out these different effects. Microturbulence is invoked often as a last resort to account for the observed, but otherwise unexplained, widths of spectral lines.

The magnetic field structure of the Sun is studied both through the coronal structure and *magnetograms* produced by polarization measuring spectrographs. Weak fields pervade the entire solar surface, but stronger concentrations are found in certain regions, especially those associated with so-called active regions, around sunspots.

Strongly correlated fluctuations on time scales of minutes and hours in the brightness and velocity of many regions over the solar disk have begun to reveal details about the interior structure of the Sun through the techniques of helioseismology.

Finally, particle emissions of the Sun are observed with detectors on satellites and probes and even indirectly on the Earth with cosmic ray detectors and through other indirect effects on the ionosphere and upper atmosphere. One particle, the neutrino, produced in nuclear reactions in the core, has been of great significance in the past few decades. The neutrino observations revealed a shortfall of about a factor of 3 in neutrino flux produced from expected nuclear reactions in the proton–proton cycle that is the expected source of solar energy. Until recently, this neutrino problem was one of the major unsolved problems of astrophysics. The solution lay in the oscillation between different forms of the neutrino, from that related to the electron to those related to the tau meson and the mu meson (the tauon and muon, respectively). Earlier experiments were sensitive only to the electron neutrino, and so failed to see the number that converted to the other "flavors" by the time they reached the detectors. The critical evidence for neutrino oscillation is presented in Ahmad et al. (2001) and a recent examination of the agreement among solar models, solar seismomology observations, and neutrino data can be found in Bahcall et al. (2005).

4.4 The Sun's Radiation

In order to discuss the radiative properties of the Sun, a few general ideas, definitions, and distinctions need to be considered.

4.4.1 Luminosity and Surface Brightness

In spherically symmetric stars, the radiant power emitted to the rest of the universe (the *luminosity*) is given by

$$\mathcal{L} = 4\pi\mathfrak{R}^2\mathfrak{S} \quad \text{(watts or W)} \tag{4.6}$$

where $\mathfrak{R}$ is the stellar radius and $\mathfrak{S}$ (a script S) is the power radiated per unit surface area. In some branches of astronomy, $\mathfrak{S}$ [sometimes written as an unscripted S, a script B, σ, Σ, or, because it has units of flux, as $\mathcal{F}$] is called the *surface flux*, or, sometimes, the *surface brightness*. It may also be called the *radiance*, which Dufay (1964, p. 5) defines as "the flux radiated by a unit surface element in all exterior directions."[2] Most often, the quantity referred to as the surface brightness or radiance is expressed in units of power per unit area per unit solid angle of emission (see further discussion below). With these units, these quantities measure *intensity*. The distinction is important and will be discussed further in this section and in Section 4.4.3.

In (4.6), there is no explicit dependence on solid angle and radiation has been integrated over the entire sphere. In any case, $\mathfrak{S}$ depends on the temperature, which is basically a measure of the average kinetic energy of a large number of particles. $\mathfrak{S}$ may be written

$$\mathfrak{S} = \sigma T^4 \qquad (\mathrm{W/m^2}) \tag{4.7}$$

where σ is the *Stefan–Boltzmann constant*, $(5.6705\pm2)\times10^{-8}\ \mathrm{Wm^{-2}K^{-4}}$, and T is called the *effective temperature*—the temperature at the star's surface in order to have a surface brightness equal to that of the idealized, perfect radiator that we commonly call a *black body* (BB). Indeed, the surface of a star is *not* a black body, as comparisons between BB curves for particular temperatures and real star's radiation curve tracings readily reveal (see Figure 4.1). However, for many purposes, especially in the interiors of stars, BB approximations are very useful.

Before we discuss black body radiation, however, we must review the physical terms we have just used and discuss how these terms fit into astronomy, and the more specific fields of practical physics called radiometry and photometry.

The *luminosity*, defined as the amount of energy radiated per second at the source (in the cases described here, the Sun or a star) is also called the *radiant power* or, sometimes, regrettably, "*radiant flux*" (in units of W). The power radiated only in a spectral region centered around the wavelength $0.555\,\mu$m (555 nm, 5550), which is the approximate wavelength of peak sensitivity of the human eye in daylight, is referred to as *luminous power*, or, sometimes as "*luminous flux*." The unit of luminous power is, naturally enough, the *lumen* (abbreviated lm), equivalent to[3] 1/683 W. *Radiant exitance* (*luminous exitance* for the visual region) is the power emitted per unit area in units of W/m^2 (and in the visual, lm/m^2); *radiant intensity* (*luminous intensity* for the visual region) is the amount of *radiant* power (*luminous* power in

[2] If one is discussing the measurement of light from a source, *irradiance* is a term sometimes used to describe the radiation in a wide passband that falls onto a surface (of a telescope, or a detector, for instance); if the measurement involves such received radiation per unit of wavelength, it is called *spectral irradiance*.

[3] See Meyer-Arendt (1995, p. 351); the sensitivity of the human eye to different wavelengths makes this conversion factor vary somewhat with wavelength and bandwidth.

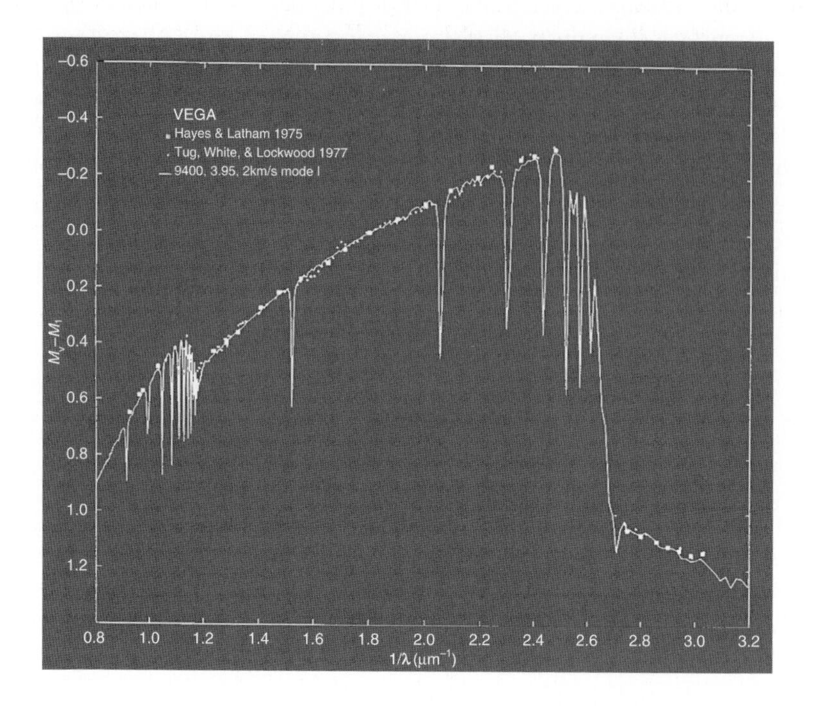

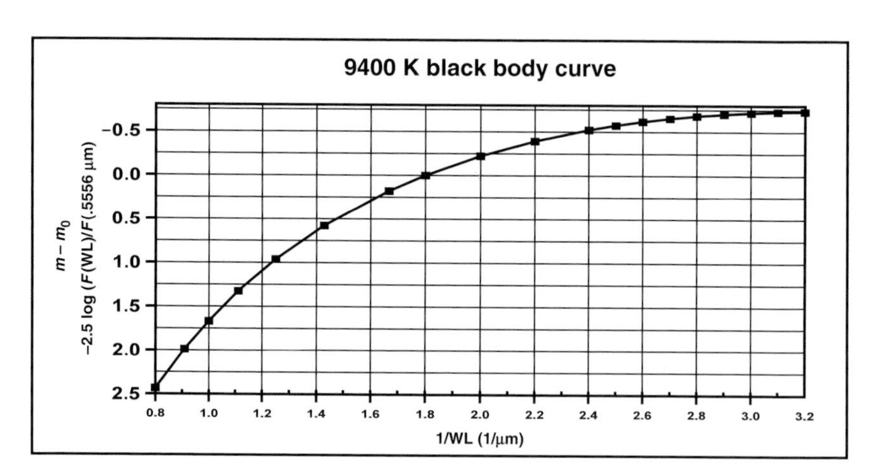

Fig. 4.1. **a** From Kurucz (1993), the spectral irradiance of Vega, relative to its value at 5556 and converted to magnitudes, compared to Kurucz's model and others. **b** The monochromatic flux of a 9400 K black body, the effective temperature of that model, and normalized and converted to magnitudes as in **a**. The abscissa is in units of inverse wavelength, so that a wavelength of 0.5 μm is located at 2.0 $(\mu\text{m})^{-1}$

the visual) emitted per unit solid angle, Ω, and has units of W/sr (for the visual region, this is in units of lm/sr). If a source emitting monochromatic radiation at a frequency of 540×10^{12} Hz (i.e., at a wavelength of 555 nm) into a given direction has a radiant intensity of (1/683)W in that direction, the luminous intensity would be 1 lm/sr, a quantity also called a *candela* (cd). Finally, the amount of radiation through a unit area and unit solid angle[4] at the source is the *radiance*; *luminance* refers to the visual component of the radiance. Radiance and luminance are used to describe the power emitted at different regions of the emitter's surface, and they are sometimes referred to as "brightness," or "surface brightness." The units are $\mathrm{W m^{-2} sr^{-1}}$ for radiance, and $\mathrm{cd/m^2}$ [or *nit* (for the Latin *nitere*, "to shine")], for luminance. Alternatively the luminance is given in units of lamberts ($10^4/\pi\,\mathrm{cd/m^2}$). In astronomy, the relative brightness is often expressed in magnitudes, related to the various units of brightness through the base 10 logarithm of the brightness ratio: $(m - m_0) = -2.5 \log (\ell/\ell_0)$. The only requirement is that ℓ and ℓ_0 must be measured in the same units. We will discuss magnitudes and their uses in a later section and in later chapters.

In Figure 4.1, we show the monochromatic (i.e., per unit wavelength) flux in the form of spectral irradiance of the star Vega, along with predicted values from several models (cited in Kurucz, 1993) and, separately, in the same units, the corresponding radiation curve of a black body at 9400 K, the temperature assumed for the model. The spectral irradiance is normalized to (i.e., divided by) its value at 555.6 nm equivalent to the inverse wavelength, $1.8\,(\mu\mathrm{m})^{-1}$, against which units the spectra are plotted. Vega's spectrum shows absorption features at individual wavelengths (*called spectral lines*) and in the continuum. The image is reversed to white on black to emphasize this point.

The role of opacity in shaping spectral lines will be featured in Section 4.8. Absorptions over a large wavelength range are due to *bound–free* and *free–free transitions*. The former can be caused by ionizations of atoms by photons with energies greater than the ionization energy of the atom. The edge of the Balmer series of hydrogen, caused by ionization of the hydrogen atom with electrons in the second energy level, is seen at $\sim$0.365 μm equivalent to $2.74\,(\mu\mathrm{m})^{-1}$. Radiation streaming through the star's atmosphere that is shorter in wavelength (bluer) than the *Balmer limit* is readily absorbed by atoms in this energy state, so that the outward going flux is greatly reduced.

---

[4] Angles are measured in degrees or radians (2π radians $= 360°$). Solid angles are measured in square degrees or steradians (sr). Generally, $\Omega = \mathrm{area/(distance)^2}$. The surface area of a sphere of radius R meters is $4\pi R^2\,\mathrm{m}^2$, so that from the center, $\Omega = 4\pi$ sr; 1 steradian is the angle subtended by an area of one square meter on the surface of a sphere of 1 m radius (NB: the area can be of any shape). Also, $(\pi/180)^2 = 3.0462 \times 10^{-3}\,\mathrm{sr/deg^2}$ and $1\,\mathrm{sr} = (180/\pi)^2 = (57.296)^2 = 3282.8\,\mathrm{deg^2}$; the entire sphere subtends at the center $4\pi\,\mathrm{sr} = 41,252.88\,\mathrm{deg^2}$.

4.4.2 Black Body Radiation

Black bodies are characterized by the following radiation properties:

1. The distribution of light with frequency or wavelength (henceforth: f/wl) is called the *Planck function* (or sometimes the *Kirchhoff–Planck function*):

$$\begin{aligned} f_\nu(T) &= (2\pi h\nu^3/c^2)(\mathrm{e}^{h\nu/kT}-1)^{-1}\ (\mathrm{Wm^{-2}\,Hz^{-1}}) \\ f_\lambda(T) &= (2\pi hc^2/\lambda^5)(\mathrm{e}^{hc/\lambda kT}-1)^{-1}\ (\mathrm{W/m^2/wl\,unit}) \end{aligned} \tag{4.8}$$

   where Planck's constant, $h = 6.6261 \times 10^{-34}$ J $\cdot$ s, Boltzmann's constant, $k = 1.3807 \times 10^{-23}$ J/K, and the speed of light, $c = 2.9979 \times 10^8$ m/s.

2. The setting of the derivative of (4.8) to zero defines the peak of the curve, and the f/wl at which this occurs is related to T through:

$$\begin{aligned} \nu_{\mathrm{max}} &= 0.5879\,\mathrm{T}\ (\mathrm{Hz}) \\ \lambda_{\mathrm{max}} &= 0.002898/\mathrm{T}\ (\mathrm{m}) \end{aligned} \tag{4.9}$$

   The latter expression is known as *Wien's law.*

3. The integrating[5] of equation (4.3) over all f/wl gives the *black body emittance*, the total flow of radiation outward from a black body surface per unit area:

$$\mathcal{F} = \sigma T^4\ (\mathrm{W/m^2}) \tag{4.10}$$

   where $\sigma \equiv (\pi^2/60)\,k^4/[(h/2\pi)^3c^2] = 5.6705\times10^{-8}\,\mathrm{Wm^{-2}\,K^{-4}}$ and T is not only the effective temperature, but also the local temperature as measured at any point on the black body (a consequence of the property known as *local thermodynamic equilibrium* (LTE)—more on this in Section 4.9). Equation (4.10) is called the *Stefan–Boltzmann law*, and so is (4.7) because the temperature there is understood to be an *effective* temperature.

Example 4.1

Figure 4.2 is a plot of the base 10 logarithm of the Planck function for a 10,000 K black body, computed from the second part of (4.8) against the base 10 logarithm of the wavelength. The peak of the curve, computed from the second part of equation (4.9), is: $\lambda_{\mathrm{max}} = 2898/T\,(\mu\mathrm{m}) = 0.2898\,\mu\mathrm{m} = 289.8\,\mathrm{nm}$, in the UV.

---

[5] The process of integration can be thought of as the summing of many products of the form $f_\nu\,\mathrm{d}\nu$, where f_ν is the mean of $f_\nu(\nu)$ and $f_\nu(\nu + \mathrm{d}\nu)$, and $\mathrm{d}\nu$ is a small difference in frequency. The sum is taken over a range of frequency or, for the analogous summing of products involving f_λ, over a range of wavelength.

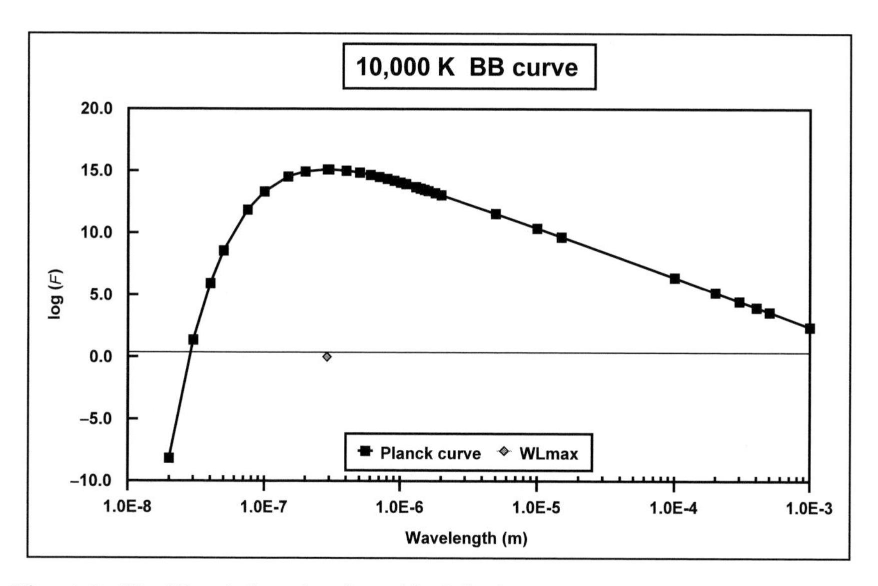

Fig. 4.2. The Planck function for a black body at $T = 10,000$ K. Note that the peak of the radiation curve occurs near $\lambda = 3 \times 10^{-7}$ m, marked by the lozenge on the line parallel to the x-axis

Example 4.2

The black body curves computed for different temperatures are nested: lower temperature curves fall completely below higher temperature curves at each and every f/wl. Figure 4.3 illustrates this for Plank functions of two different temperatures, 1000 and 300 K, respectively. This means that the flux integrated across all f/wl will also be greater for a higher temperature black body, as predicted by the Stefan–Boltzmann law (4.10). Note also that the higher temperature black body would appear bluer because the peak of its radiation curve occurs at shorter wavelengths.

4.4.3 Flux and Intensity

The quantity B_ν has units of intensity and, in the present context, is sometimes called the *source function*. Allen (1973, p. 104) defines the integration of this quantity over all f/wl as:

$$B = (\sigma/\pi)\, T^4 = 1.80468 \times 10^{-5}\, T^4 \,(\text{erg cm}^{-2}\,\text{s}^{-1}\,\text{sr}^{-1}) \qquad (4.11)$$

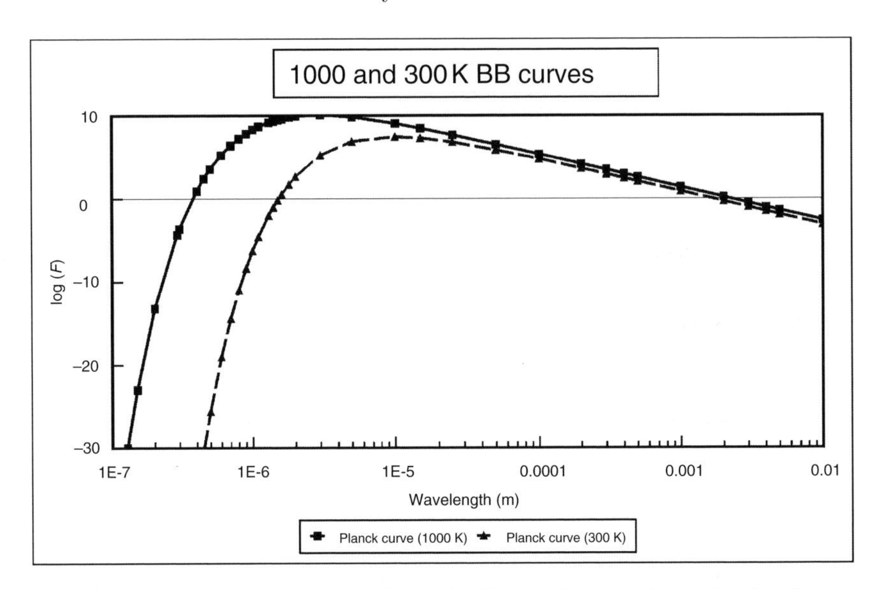

Fig. 4.3. The Planck functions for black bodies at $T = 1\,000$ and 300 K. Note the shift to the blue of the wavelength of the peak of the higher temperature–radiation curve, and that the lower temperature curve completely nests within the higher temperature curve

evaluated in CGS units, i.e., B is the emittance per unit solid angle (steradian). Note that the quantity, that we have called $\mathfrak{B}$ in (4.6) and (4.7) (and written as a script B sometimes in other sources), has units of power per unit area, whereas the B of (4.11) has units of power per unit area per steradian of solid angle. These are the units of *intensity*.

Intensity is a directed flux of energy, whereas emittance, surface brightness (as we have referred to it here), or more commonly, *flux*, is the intensity integrated over all outgoing (increasing r) directions:

$$f = \int B \, \mathrm{d}\omega = 2\pi B \int_0^{\pi/2} \sin\theta \ \cos\theta \ \mathrm{d}\theta \tag{4.12}$$

where θ is the angle with respect to the surface normal of a particular beam of radiation. Figure 4.4 illustrates the geometry. The important distinction between flux and intensity is that the intensity depends on the angle of emergence from a star. At the center of a stellar disk the intensity is I_0, and at some point where the line of sight is at an angle θ, the intensity is often given as: $I = I_0\ (1 - u + u \cos\theta)$. See Sections 4.5.1 and 4.5.2 for further discussion of *limb-darkening*. In Section 4.8 we discuss the transport of radiation through material with a "transfer equation" and use it to explain how emission and absorption features arise in the spectra.

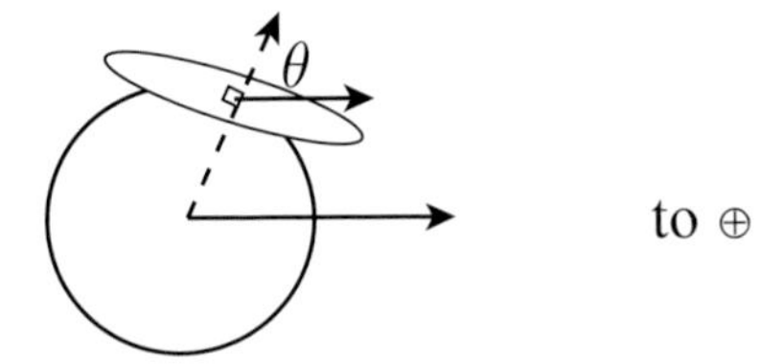

Fig. 4.4. Directed beams of radiative intensity

4.4.4 Observed Radiative Properties of the Sun

The *solar constant* is the flux of total radiation received outside the Earth's atmosphere per unit area at the mean distance from the Sun. It is equal to

$$\begin{aligned}\mathfrak{C} &= 1.950(4)\,\mathrm{cal/cm^2\ min} = 1.360\times10^{6}\,\mathrm{erg\,cm^{-2}\,s^{-1}}\\ &= 1.360\times10^{3}\,\mathrm{W/m^2},\end{aligned}$$

where we have used a script C to represent the solar constant. The luminosity can then be determined by multiplying the solar constant by the total surface area of a sphere of radius 1 au:

$$\begin{aligned}\mathcal{L}_\odot &= 4\pi r^2\,\mathfrak{C} \\ &= 3.826(8)\times10^{33}\,\mathrm{erg/s} = 3.826(8)\times10^{26}\,\mathrm{W}\end{aligned} \tag{4.13}$$

This is one way to determine the luminosity. Another way is to multiply the surface area of the Sun by its surface brightness. From these data, the mean solar surface brightness and the mean solar intensity can be determined:

$$\mathcal{F} = \mathbf{S} = 6.27\times10^{7}\,\mathrm{W/m^2} = 6.27\times10^{10}\,\mathrm{erg\,cm^{-2}\,s^{-1}}$$

$$I = \mathcal{F}/\pi \tag{4.14}$$

$$\text{So,}\ \ I = 2.000\times10^{10}\,\mathrm{erg\,cm^{-2}\,s^{-1}\,sr^{-1}} = 2.000\times10^{7}\,\mathrm{Wm^{-2}\,sr^{-1}}$$

Finally, the effective temperature can be computed. One such determination is $T_{\mathrm{eff}} = 5777\,\mathrm{K}$ (W. C. Livingston, in Cox 2000, p. 341).

The observed visual magnitude of the Sun is $V = -26.75$ and its color index (B-V) = 0.65. The absolute magnitude is the magnitude of a star as it would appear at a distance of 10 pc. Therefore,

$$M = m - 5\ \log(r/10) \tag{4.15}$$

where r is expressed in parsecs. From (4.15), $M_{\odot V} = 4.82$. The *bolometric correction* (BC) is the difference between the visual and bolometric magnitudes. Defined as in (4.16), if non-zero, it is always *negative*:

$$M_{\mathrm{bol}} = M_V + \mathrm{BC} \tag{4.16}$$

The bolometric correction for the Sun is −0.08 so its bolometric absolute magnitude is 4.74. Slightly different values may be obtained elsewhere in the literature; we used the data given in Allen's Astrophysical Quantities (Cox 2000).

4.5 The Photosphere

The visible surface of the Sun is known as the photosphere (see Figure 4.5). We see it through the overlying chromosphere and corona. The total thickness of the photosphere is only a few hundred kilometers, yet most of the visible radiation comes from this region. The reason for this interesting circumstance can be found in an understanding of opacity and limb-darkening, which we discuss in the following section. The photosphere marks the upper end of the convection zone, the convective effects of which are manifested in granulation

Fig. 4.5. The photospheric disk of the partially eclipsed Sun, February 26, 1979, through an Hα filter, showing limb-darkening and sunspots. Courtesy Dr T. A. Clark

patterns on the photosphere. We will discuss first the visibility of radiation and the causes of limb-darkening, then the convection zone, and finally other features of the photosphere.

4.5.1 Opacity and Optical Depth

We start by defining the unitless *optical depth*, a measure of the extent to which radiation can penetrate into any medium. The contribution to the optical depth of a cylinder of gas of physical length dx, cross-section per unit mass or mass absorption coefficient[6] k_λ (expressed in units of m^2/kg), and density ρ (in units of $\mathrm{kg/m}^3$) is:

$$\mathrm{d}\tau_\lambda = k_\lambda \rho \, \mathrm{d}x \tag{4.17}$$

Integrating over a total physical distance (X–0) or X, we get:

$$\tau_\lambda = \int_0^X k_\lambda \rho \, \mathrm{d}x \tag{4.18}$$

When $k_\lambda \rho$ is large and/or the physical length is great, τ_λ is also large and so is the absorption. Note that it is the product of the physical length and the absorption coefficient $k_\lambda \rho$ that determines the optical depth and thus the absorption. A very long path length need not guarantee strong absorption (or emission) if the opacity or the density is very small. If the density and opacity are constant over this total length, one could write

$$\tau_\lambda = k_\lambda \rho (X - 0) = k_\lambda \rho X \tag{4.19}$$

When radiation travels radially outward through a star, the intensity of the radiation changes over any infinitesimal distance, dr, due to the opacity of the material through which it travels

$$\mathrm{d}I = -I k_\lambda \rho \, \mathrm{d}r = -I \, \mathrm{d}\tau_\lambda \tag{4.20}$$

where I is intensity at position r. The negative sign indicates a loss of intensity. Integrating over the optical depth, we can express the intensity, I, emerging

[6] Note that the *absorption coefficient* discussed in, e.g., Schlosser et al. (1991/4) is in units of inverse length, m^{-1}, not in m^2/kg, the units for the mass absorption coefficient, k_λ. This is because we show the density dependence explicitly here; but in (4.32), Section 4.8, we define and use the absorption coefficient. Generally, the absorption coefficient is referred to as *opacity*.

from a column somewhere on the disk of a star, in terms of that at the base of the column, by

$$I = I_0 \exp(-\tau_\lambda) \tag{4.21}$$

Of course radiation is not only absorbed and scattered out of the beam, but can be scattered and reemitted into it also. We can express the radiation injected into the beam by the term *emissivity*, a quantity that is proportional to the mass absorption coefficient. The emissivity of the solar/stellar atmosphere at any depth can be written:

$$\varepsilon_\lambda = \int B_\lambda(T) k_\lambda \rho \, \mathrm{d}r = \int B_\lambda(\tau_\lambda) \, \mathrm{d}\tau_\lambda \tag{4.22}$$

B_λ, r, τ_λ, and ε_λ increase *inward* (i.e., into the solar atmosphere from outside). Upwelling radiation that starts out being well represented by a Planckian will become absorbed due to the presence of absorbers in the line of sight. This can be represented by an equation of the kind:

$$I_\lambda = \int_0^\infty B_\lambda(\tau_\lambda) \, e^{-\tau_\lambda} \, \mathrm{d}\tau_\lambda \tag{4.23}$$

Along any particular ray, an optical depth of unity corresponds to the depth over which the radiation falls to $1/e$ of its original value. From (4.21),

$$\tau_\lambda = 1 \tag{4.24}$$

From the outside, we see little radiation from depths deeper than $\tau_\lambda = 1$. Therefore, this is a typical depth to which we can see into the atmosphere along any line of sight.

Note from Figure 4.6 that from a given physical depth, say point P′, measured along a radius, the path length along a normal to a point on the surface is shorter than the path length through the solar atmosphere to the direction of the observer, unless the beam is emerging from the disk center. Thus the radiation is diminished over this "slant ray" distance compared to one traversing the "normal ray" distance. However, essentially we do not "see" deeper than $\tau = 1$ along any ray.

4.5.2 Center-to-Limb Variation

For this section, we define r as a depth variable (r = 0 at the stellar surface). Following our previous discussion, along a slant ray (r' direction) from some given source, angled by θ with respect to the surface normal (r direction) the physical depth is:

$$\mathrm{d}r' \approx \mathrm{d}r / \cos\theta = \sec\theta \, \mathrm{d}r \tag{4.25}$$

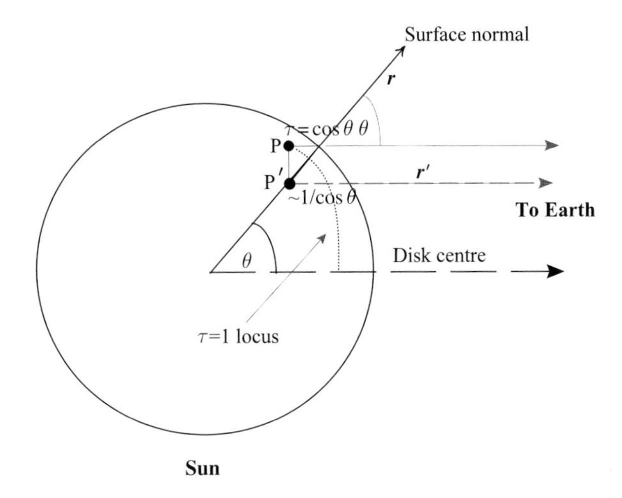

Fig. 4.6. Solar limb-darkening definitions and geometry

(see Figure 4.6). For the optical depth along this slant ray, we can write,

$$\tau_\lambda{}' = \int_{r=0}^{r=r'} k_\lambda(r)\rho \, \mathrm{d}r' = \int_{r=0}^{r=r'} k_\lambda(r)\rho \, \sec\theta \, \mathrm{d}r = 1 \tag{4.26}$$

where $r' \geq r$. It is set equal to 1 because this is the optical depth to which we "see" along the slant ray (θ).

The optical depth corresponding to the physical depth r, at point P, say, is given by

$$\tau_\lambda = \int_{r=0}^{r=r} k_\lambda(r)\rho \, \mathrm{d}r = (1/\sec\theta) \int_{r=0}^{r=r'} k_\lambda(r)\rho \, \sec\theta \, \mathrm{d}r \tag{4.27}$$

which then simplifies to: $\tau_\lambda = (1/\sec\theta) \cdot 1$, so that at this θ, we actually see down to an optical depth (measured along the normal ray) of only:

$$\tau_\lambda = \cos\theta \tag{4.28}$$

Thus, by examining the intensity emerging from a star at a range of angles θ from the normal (i.e., from the direction of the center of the disk), a table of τ_λ vs. I_λ can be obtained.

This is the *center-to-limb variation* or *limb-darkening* which can be seen in images of the Sun's disk. The limb-darkening is usually defined by the ratio

$$I_\lambda(\theta)/I_\lambda(0)$$

where $I_\lambda(\theta)$ is the intensity of the solar radiation arising from a point on the solar disk which makes an angle θ measured between a line from the center of the Sun to the point and a line from the Sun's center to the observer (as in Figure 4.6).

The limb-darkening is often expressed analytically in the linear (but not always most accurate) form,

$$\begin{aligned} I_\lambda(\theta)/I_\lambda(0) &= 1 - u + u\cos\theta \\ &= 1 + u\,(\cos\theta - 1) \end{aligned} \tag{4.29}$$

where u is known as the *linear limb-darkening coefficient*. Therefore, for a fully darkened limb, $u = 1$; for no limb-darkening, $u = 0$. See Schlosser et al. (1991/4, Fig. 28.7) for a plot of $I(\theta)/I(0)$ for two passbands and for a plot of T vs. r (Fig. 28.8) in the Sun. Limb-darkening can be seen even in the darkened limb of white light images of the Sun. It varies slowly with wavelength in the continuum, but generally differs from one spectral line to another, because of the different opacities per absorbing element and the different conditions at the atmospheric heights where the atoms contributing to the line are located. Figure 4.7 shows the intensities of a star similar in temperature to, but much larger than, the Sun at different limb positions ($\mu = \cos\theta$). The significance of limb-darkening lies in the circumstance that the temperature increases with depth into the star.

Assuming a value of T_0, the temperature at the center of the disk (not at the center of the Sun!), one can then determine a series of values of T vs. τ and

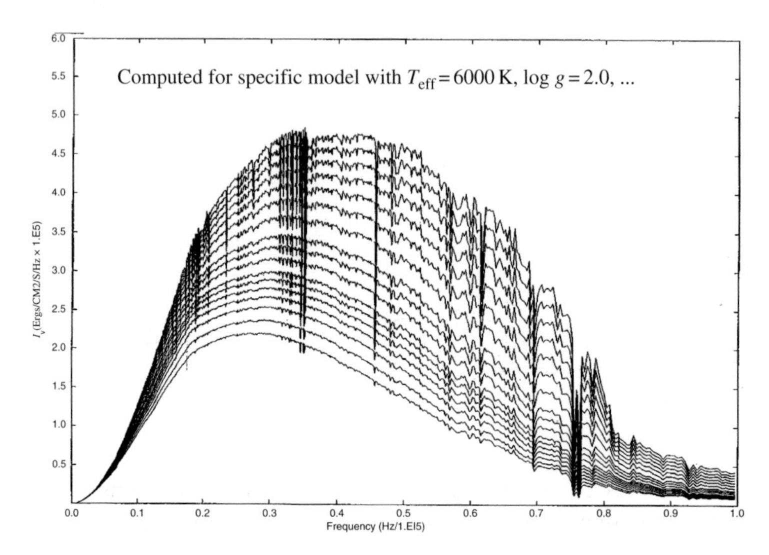

Fig. 4.7. Models of limb-darkening: computed stellar intensity for values of $\mu = \cos\theta = 1$, 0.9, 0.8, 0.7, 0.6, 0.5, 0.4, 0.3, 0.25, 0.2, 0.15, 0.1, 0.125, 0.1, 0.075, 0.05, 0.025, and 0.01, for images top to bottom, respectively (from Kurucz 1993)

thus provide information about the thermal structure of the solar atmosphere. For the Sun, $I_0 = 2.41 \times 10^7 (\text{Wm}^{-2}\text{sr}^{-1})$, which yields a disk-center temperature, $T_0 \approx 6050\,\text{K}$.

Note that if the temperature were to decrease with depth, we would expect (depending on the behavior of the opacity) greater emission from the limbs than from the disk center. This is known as *limb-brightening*, and this is actually seen in the Sun at some wavelengths where the emission arises in the Sun's upper atmospheric levels. The more general term, therefore, is *center-to-limb variation.*

In the solar photosphere, the temperature increases inward. The temperature minimum is reached in the photosphere at the base of the chromosphere, above which it climbs to very high values.

In the ultraviolet, the center-to-limb variation shows more complicated behavior than the linear form of (4.29) provides (refer to the SkyLab satellite analyses of solar UV center-to-limb variation by Kjeldseth Moe and Milone 1978). This reflects the effect of the temperature reversal in the chromosphere and the possibility of contributions of atoms and ions from more than one level of the solar atmosphere.

From eclipsing binary star light curves, limb-darkening can also be measured on the disks of other stars. Equation (4.29) is most often used in this context, but even then, non-linear limb-darkening forms are sometimes found necessary to achieve satisfactory fitting of theoretical light curves to the branches of the light curve minima. Modern light curve analysis programs now have square-root and logarithmic as well as linear forms of the limb-darkening law (see Kallrath and Milone 1999, pp. 95–98).

4.5.3 Granulation and the Convection Zone

The photosphere lies atop an extensive *convection zone*, in which rising parcels of gas expand and cool and descending parcels are compressed and become hotter. Observationally, the convection zone shows up as large cells of upwelling material about 10^3 km across known as *granules*. The upward speeds of the hot gas in these granules is $\sim$500 km/s. This gas radiates as it nears the surface, cools, and sinks back down at the edges of the convective cells. These narrow cell boundaries appear darker by contrast. Larger regions of organized motions called *supergranules*, up to 3×10^4 km across, underlie the smaller cells. Although the upper velocities of these motions are smaller, they manage to concentrate magnetic fields in the lower chromosphere into networks, seen as heated regions in strong-line spectroheliograms.

Convection arises if the change in pressure with temperature through the stellar atmosphere is larger than would occur in a parcel of gas that rises *adiabatically*, i.e, without heat loss. Such an effect can occur when there is strongly absorbing material. At the core of the Sun, the temperature is expected to be $\sim 15 \times 10^6$ K, and the energy created through nuclear reactions is radiated away from the inner $0.25\Re_{\odot}$. At a distance of $\sim 0.71\Re_{\odot}$ from the center, the temperature has dropped to $\sim 2 \times 10^6$ K and the opacity of the matter to radiation begins to increase as highly ionized atoms begin to recapture electrons, and so are eligible to absorb photons. Whereas recapture results in photon emission, outgoing radiation is depleted as the absorbed radiation is reradiated in all directions. This increased opacity deep in the envelope of the Sun triggers the start of the solar convection zone.

4.5.4 Other Photospheric Features

Above the outer edge of the convection zone, one of the principal sources of opacity in the solar photosphere is the absorption of radiation by the H^- ion, a hydrogen atom to which a second electron has been loosely attached. This was first suggested as a major source of the continuous opacity in the optical spectrum of the Sun by Wildt (1939).

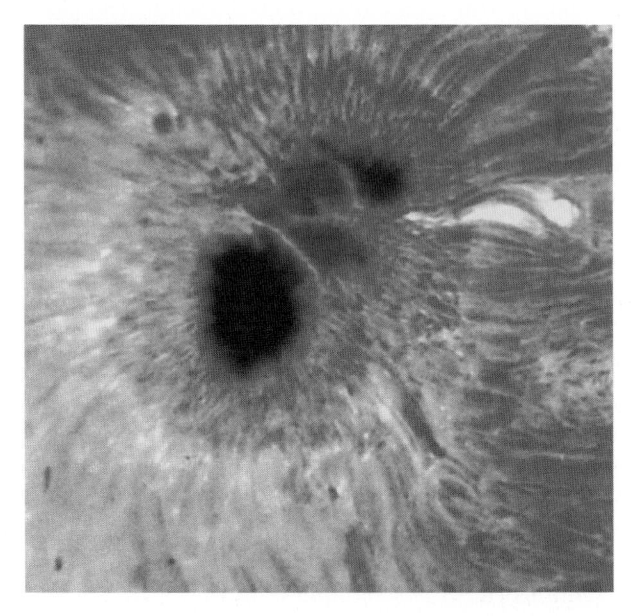

Fig. 4.8. An active sunspot group (AR 8971) as recorded in the Hα spectral line on April 27, 2000, with the McMath–Pierce Solar Telescope at the Kitt Peak National Observatory, Arizona. The area covered is 100×100 arc-secs. Courtesy Dr T. A. Clark. The McMath–Pierce Solar Telescope is operated for the National Science Foundation by the Association of Universities for Research in Astronomy as part of the National Solar Observatory (See also Plate 1)

Magnetic fields are observed on the face of the Sun; they probably arise from the convective motions of charged particles and become amplified close to the surface. The overall field of the Sun is relatively weak: $\sim 10^{-4}$ T. This can be compared to the value at the Earth's surface, $\sim 4 \times 10^{-5}$ T. From time to time, localized regions of strong magnetic fields appear at low to mid-latitudes. These fields inhibit convection and result in cooler and thus darker areas than the surrounding photosphere. We know these darker areas as sunspots. The numbers of, and total area occupied by, sunspots vary with a period of about 11 years.

The image of a sunspot group in the light of the core of the Hα line in Figure 4.8 shows the detailed structure of the Sun in the vicinity of the spots, and demonstrates the strong influence of magnetic fields in shaping the structure. Active regions are given numerical designations; that shown is called AR8971. A flare can be seen in progress on the right; plages or faculae, brightenings around the spots, in discrete spectral lines and in white light, respectively, can also be seen. Such events are numerous near sunspot cycle maximum.

4.6 The Chromosphere

The temperature of the solar atmosphere decreases with increasing radius through the photosphere and reaches a minimum ($\sim$4200 K) in the lower chromosphere. It then rises to an intermediate plateau before increasing rapidly in a *transition zone*, eventually reaching millions of degrees in the solar corona.

Different regions of the solar atmosphere can be studied by observations at different wavelengths within spectral line profiles. Strong lines saturate quickly in the line center and thus represent the highest level of the solar atmosphere. In the "wings" of the line, the physical source depth is greater; the radiation comes from deeper layers of the atmosphere. Ultraviolet, x-ray, and radio regions of the spectrum are used to explore the outer solar atmosphere: the chromosphere, the transition region, and the corona. The chromosphere is the source of most of the ultraviolet radiation that impacts the upper atmosphere of the Earth.

The chromosphere extends upward from the photosphere for $\sim$2000 km. The gas density drops to $\sim 10^{-4}$ that of the photosphere. Although it is 40$\times$ thicker, this layer has much weaker absorption and thus lower optical depth over most of the visible spectrum, so that it appears only $\sim 10^{-4}$ as intense as the photosphere. The temperature varies only slightly with height, from a minimum of $\sim$4400 to 25,000 K, over most of the chromosphere. Beginning

~2000 km above the photosphere, the density decreases rapidly by ~3 orders of magnitude and the temperature rises sharply to $\sim 10^6$ degrees.

Spectroheliograms taken in the cores of strong spectral lines, such as Hα, reveal the structure of the chromosphere and clearly show the presence of a *chromospheric network*, a region of concentrated magnetic fields that contain the enigmatic *spicules*. The fine structure in an active region (AR10812) in Hα can be seen in Figure 4.9, from Fig. 1b of van Noort and Ruppe van der Voort (2006). Plages, pores, surges, and spicules are visible.

The chromosphere (and at very short wavelengths, the lower corona) is revealed also in the spectral lines of the far UV, where the opacity is also high. At wavelengths below about 1600 (160 nm), the solar spectrum changes from an absorption to an emission spectrum, indicating the absence of cooler radiation layers above the emitting region. Figure 4.10 displays the spectrum of the Sun in the "rocket ultraviolet," so-called because the Earth's atmosphere will not pass radiation at wavelengths shorter than ~320 nm. The echelle spectrograph used to obtain this spectrum was carried aloft on

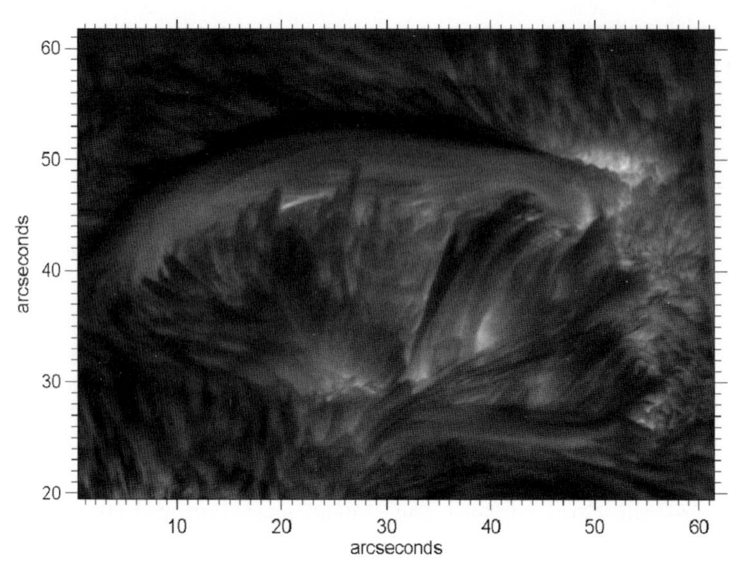

Fig. 4.9. The decaying active region AR10812, as seen on October 4, 2005 in an image taken with the Swedish 1-m Solar Telescope by L. H. M. Rouppe van der Voort and M. J. van Noort, University of Oslo. A time series made of many carefully aligned images of this and another active region revealed movement of bright points with horizontal speeds of between 100 and 240 km/s, and somewhat slower features moving in loops. A bright surge is seen in the *top right part* of the image. Flame-like spicules are also visible. Courtesy of van Noort and Rouppe van der Voort (2006, Fig. 1), and reproduced by permission of the AAS and the authors

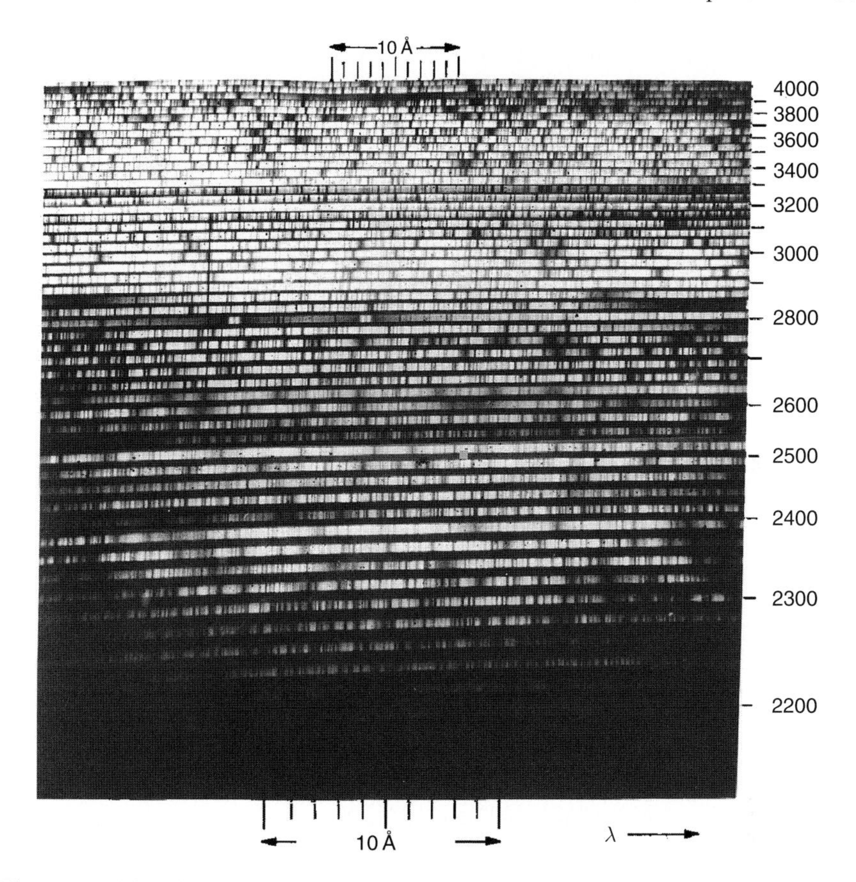

Fig. 4.10. The ultraviolet spectrum of the center third of the solar disk, obtained with an echelle spectrograph on an Aerobee rocket in 1961. The prominent bifurcated emission features superimposed on broad absorption lines near 2800 are the h and k lines of MG II. Courtesy Dr. R. Tousey, US Naval Research Laboratory

an Aerobee rocket to an altitude of ∼100 km above the Earth's surface. A spectral resolution as high as 0.020 was achieved (Tousey et al. 1974).

Radiation from higher levels is also seen at visual wavelengths to a certain extent, particularly in the cores of the strong absorption lines of the Balmer series of hydrogen, Ca II (H and K), and in Mg II (h and k) lines, where the emission is enhanced. Total solar eclipse observations are also very important because they provide good signal-to-noise measurements because the lower layers are masked out by the lunar disk. They can also provide high-resolution measurements, and those made from space are undisturbed by terrestrial atmospheric seeing effects.

From the SkyLab mission ∼1974, from which the solar ultraviolet limb-darkening was studied, much was learned about the structure of active regions throughout the solar atmosphere. Figure 4.11 shows the image of the Sun in

Fig. 4.11. Far ultraviolet images of the Sun from NRL's slitless spectrograph on board SkyLab. The central image is in the light of the helium II line at $\lambda = 30.4$ nm. The dark region to the lower right is a *coronal hole*, where the particles are allowed to stream away from the Sun as the solar wind (see Milone & Wilson, 2008, Chapter 11.3), resulting in a reduction in photographic density. In the *lower left*, an intense flare is so bright at its center that it has *decreased* the photographic density (a condition known as *solarization*). A second image, to the *left*, is the appearance of the Sun in the light of a transition of a highly ionized species of iron. Note that the disc of the Sun is seen only in silhouette against the glow from active regions behind the limb. Courtesy Dr. R. Tousey, US Naval Research Laboratory

the 304 line of helium II, the equivalent of the Lyman α line for ionized helium. The adjacent image is of a transition involving a highly ionized atom of iron. The solar disk is actually seen only in silhouette, outlined by active regions on the limb, and through the thin emission from the chromosphere and lower corona.

4.7 The Corona

This physically thick but extremely tenuous collection of gas, ionized particles, and dust has been likened to a thin flame. We are able to see the photosphere and even the chromosphere through it. The corona extends several solar radii out into the solar system, yet the intensity is $\sim 10^{-6}$ that of the photosphere. It is the extremely low density ($\sim 10^5$particles/cm^3) that makes this layer so transparent.

The corona has three main components: the *K-corona*, the *F-corona*, and the *E-corona.*

The K-corona is the light of the photosphere scattered by electrons in the corona. The name derives from the German *kontinuierlich*, continuous. This contribution is most important between 1 and $2.3\Re_{\odot}$, where it dominates the coronal light.

The F-corona is the contribution of scattered photospheric light by dust grains from regions beyond about $2.3\Re_{\odot}$. This scattering includes the absorption or Fraunhofer lines (so-called because Fraunhofer first detected these absorption lines in the Sun). The F-corona contributes to the *zodiacal light*, best seen after sunset in the Spring, or before sunrise in the Fall, when the ecliptic nearest the Sun stands most vertically above the horizon.

The E-corona is the source of emission lines from highly ionized atoms and is observed from all regions of the corona. The very high temperatures produce high levels of ionization and these temperatures along with the very low densities create a situation where even highly unlikely transitions can occur, giving rise to *forbidden line* emission. Normal transitions occur over intervals of 10^{-8} s. Normally particle collisions are frequent enough in gases that only these likely transitions can occur, giving rise to spectral lines of permitted transitions. Because collisions are relatively rare in the corona, radiative transitions from *metastable* levels, requiring intervals of the order of seconds, can occur.

The evidence for high temperatures in the corona comes from observations of all three coronal components. Fraunhofer lines are washed out in the K-corona, implying very high speeds of the electrons. The F-corona, scattered by much heavier dust particles, preserves the absorption lines better, but with some smearing. The emission lines of the E-corona arise from species of ions which can only appear in a very high temperature environment.

4.8 Line Absorption and Emission

The discussion in Section 4.5.1 can be expanded to show how absorption and emission lines can be understood. Beginning with equation (4.20), and writing $\kappa_\lambda = k_\lambda \rho$, for convenience, we arrive at this general form of the *transfer equation*:

$$\mathrm{d}I_\lambda/\mathrm{d}r = -I_\lambda \kappa_\lambda + \varepsilon_\lambda \tag{4.30}$$

where ϵ_λ is the emissivity per unit length. After dividing through by κ_λ, and substituting $\mathrm{d}\tau_\lambda = \kappa_\lambda\, \mathrm{d}r$, from (4.17), where τ_λ is the optical depth, we get:

$$\mathrm{d}I_\lambda/\mathrm{d}\tau_\lambda = S_\lambda - I_\lambda \tag{4.31}$$

where $S_\lambda \equiv \epsilon_\lambda/\kappa_\lambda$. Now, moving I_λ to the LHS, and multiplying both sides by $\exp(\tau_\lambda)$, we obtain

$$\exp(\tau_\lambda)\ \mathrm{d}I_\lambda/\mathrm{d}\tau_\lambda + \exp(\tau_\lambda)\ I_\lambda = \exp(\tau_\lambda)\ S_\lambda \tag{4.32}$$

Equation (4.32) can be rewritten as:

$$d/d\tau_\lambda \; [\exp(\tau_\lambda) \; I_\lambda] = \exp(\tau_\lambda) \; S_\lambda \tag{4.33}$$

Integrating over the limits $0 \rightarrow \tau_\lambda$, and assuming S_λ constant over this range, we have:

$$\exp(\tau_\lambda) \; I_\lambda - I_0 = S_\lambda \; [\exp(\tau_\lambda) - 1] \tag{4.34}$$

Dividing by $\exp(\tau_\lambda)$ we have,

$$I_\lambda = I_0 \exp(-\tau_\lambda) + S_\lambda[1 - \exp(-\tau_\lambda)] \tag{4.35}$$

where I_0 is the intensity of the beam emerging from the deeper atmosphere into the detectable region defined by $0 \rightarrow \tau_\lambda$, and S_λ is the emission added in this region. Now we can consider two special cases:

1. When $I_0 = 0$, the emerging intensity is merely:

$$I_\lambda = S_\lambda[1 - \exp(-\tau_\lambda)] \tag{4.36}$$

In local thermodynamic equilibrium (usually abbreviated to LTE), $S_\lambda = B_\lambda$. There are two further possibilities within this first case:

A. If the parcel is *optically thin*, so that $\tau_\lambda << 1$, then $\exp(-\tau_\lambda) \approx 1 - \tau_\lambda$, and

$$I_\lambda \approx S_\lambda[1 - (1 - \tau_\lambda)] = \tau_\lambda S_\lambda \tag{4.37}$$

B. If the gas is *optically thick*, so that $\tau_\lambda >> 1$, then $\exp(-\tau_\lambda) \rightarrow 0$, and

$$I_\lambda \rightarrow S_\lambda, \tag{4.38}$$

where, again, $S_\lambda = B_\lambda$, if LTE holds.

2. When $I_0 \neq 0$, for an optically thin gas parcel,

$$I_\lambda \approx I_0(1 - \tau_\lambda) + \tau_\lambda S_\lambda = I_0 + \tau_\lambda(S_\lambda - I_0) \tag{4.39}$$

When $I_0 > S_\lambda$, the second term on RHS is negative so the RHS is, in fact, less than I_0, and therefore $I_\lambda < I_0$. An absorption line results.

However, when $I_0 < S_\lambda$, we have added more emission than entered this parcel, $I_\lambda > I_0$, and an emission line is seen at this point in the spectrum.

4.9 Helioseismology

Many stars are variable in light and in mean radial velocity. Some stars are "geometric" variables, changing brightness because of eclipses by companions, or, by virtue of rotation of the star, revealing irregularities of shape (as in close binary systems) or surface brightness (starspots, faculae, etc.). Others vary because of large-scale motions of their atmospheres, as RR Lyrae, delta Scuti, or Cepheid variables.

The Sun has long been known to be variable, by virtue of its sunspots, on both short (rotation period) and long (the solar cycle) time scales in virtually all regions of the electromagnetic spectrum. Visually, this variability is at a very low level: the area occupied by sunspots very rarely exceeds 1% of the solar disk. Over the past few decades, the Sun has been shown to pulsate also (Leighton 1961; Leighton et al. 1962). These pulsations have very small amplitude, but they show up particularly well in some spectral features. These solar oscillations reveal the interior structure of the Sun. In Harvey's (1995) words, the oscillation spectrum is "rich and crowded." The process of retrieving the information is complicated and beyond the scope of our discussions. The following description of these waves and their significance can be found in more detail in Harvey (1995). The waves show up in two ways:

1. *Doppler shifts in the spectral lines.* The velocity variations are very small, $\sim 0.1\,\mathrm{m/s}$, so that very high spectral resolution is required. Only in the Sun does one have the flux to observe such a variation with precision, because one needs high signal/noise capability and very high spectral dispersion. The techniques needed to attain this are beyond the current topic of discussion, but often involve Michelson interferometry techniques. Figure 4.12 illustrates the effect on the disk of the Sun, and Figure 4.13 summarizes the results of the time-period analysis of a 3.5-day run of nearly continuous observations from Antarctica (Grec et al. 1980).
2. *Surface brightness variations.* Because the wave motion may cause material to oscillate in position, there are brightness variations accompanying the oscillation. The variation is again tiny: a few *micro*magnitudes. Stellar astronomers are doing well if they can establish light parameters to ± 0.001 magnitude. Such precision is very difficult to obtain, and only the very high flux levels of sunlight provide sufficient precision of measurement to permit it.

There are four basic types of nonmagnetic waves of general interest in the Sun. These are: acoustic waves, surface gravity waves, internal gravity waves, and Rossby waves. Only the first two produce 5-minute oscillations and are unambiguously observed. Resonant cavities within the Sun "organize" these

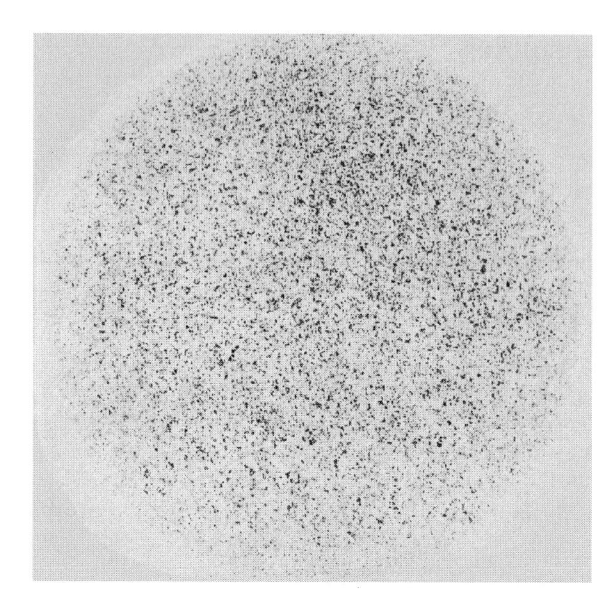

Fig. 4.12. Illustration reprinted with permission from Harvey (1995, Fig. 1). The Doppler signature of hot rising (*red*) and falling (*blue*) gas from two images obtained 2 min apart at an observatory at the South Pole in January, 1991. Copyright 1995, American Institute of Physics (See also Plate 2)

waves into standing wave patterns, which are called p-,f-, g-, and r-modes, respectively. Over ten million p- and f-modes alone have been shown to exist. The global p-modes are seen in Figure 4.13. The origin of a wave or series of waves may be the displacement of matter and a restoring force acting on the gas. In the case of acoustic waves this restoring force is the gas pressure, hence p-modes. In the case of internal gravity waves (g-modes), it is the buoyancy or vertical pressure forces. Similarly the latter serve as restoring forces for the surface gravity waves (f-modes, for *fundamental*) that propagate at the interface between the photosphere and the chromosphere.

The region which traps acoustic waves in the Sun is defined by two regimes: the lower boundary where the sound speed increases[7] and the upper boundary, near the surface, where the density decreases sharply. Figure 4.14 illustrates this. The number of radial nodes, n, characterizes the standing wave. The radial nodes are depicted in Figure 4.15, where the radial nodes run from 0 (lower right) to 40 (upper left).

[7] The result of this is to cause downward-propagating waves to bend back toward the surface.

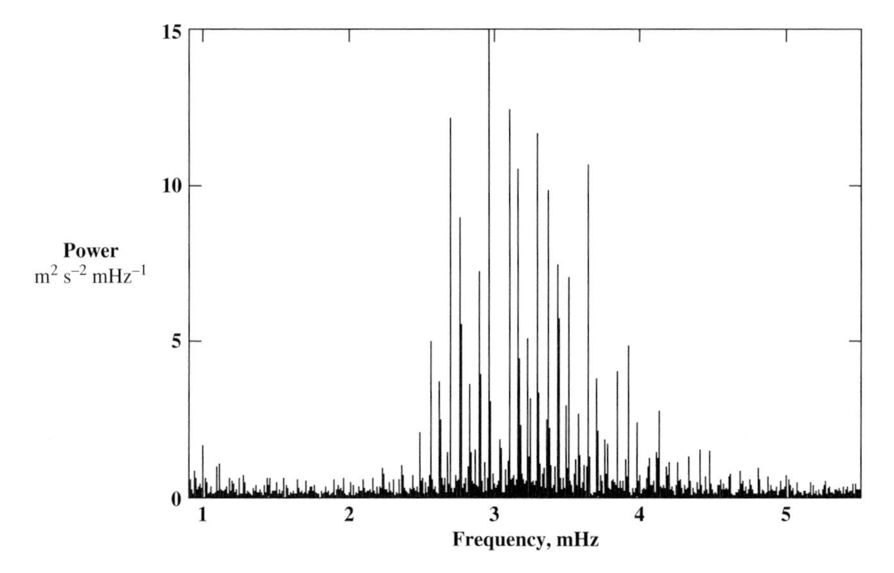

Fig. 4.13. A Doppler spectrum from the whole solar disk taken over 6 days of nearly continuous coverage. From an Antarctic observatory, showing various frequencies (inverse of periods) of gas motion. One of the strongest peaks is at ~ 0.0033 Hz, (a period of about 5 min), representing radial mode $n = 23$ and spherical harmonic $\ell = 0$. Adapted from Grec et al. (1983).

The standing waves on the Sun's circumference can be seen on great circles such as longitude circles and on small circles, such as latitude circles. These are characterized by the spherical harmonic, ℓ, and by the azimuthal order, m. The depth of the 'cavity' is revealed by the quantity ℓ/f, where f is the cyclic frequency, in the sense that the smaller this ratio, the deeper the cavity. The expected wavelengths, ℓ_n, and frequencies, f_ℓ, of acoustic waves are determined by the radial distance, r, of the cavity bottom from the Sun's center, and by ℓ, the nodal lines (circles):

$$\lambda_n = (2\pi r)/\sqrt{\ell(1+\ell)} \tag{4.40}$$

or

$$\mathrm{f}_\ell = (2\pi)/t_{\lambda n} = (2\pi\ v_{\mathrm{s}})/[(2\pi\, r)/\sqrt{\ell(\ell+1)}] \tag{4.41}$$

$$= (\gamma P/\rho)^{1/2}[\ell(\ell+1)]^{1/2}/r \tag{4.42}$$

where $\mathrm{t}_{\lambda n}$ is the period, v_{s} is the speed of sound, γ is the ratio of specific heats, P is the pressure, and ρ the density (see Milone & Wilson, 2008, Chapter 10.2 for definitions and discussion). As

$$v_{\mathrm{s}} \propto \sqrt{T}, \tag{4.43}$$

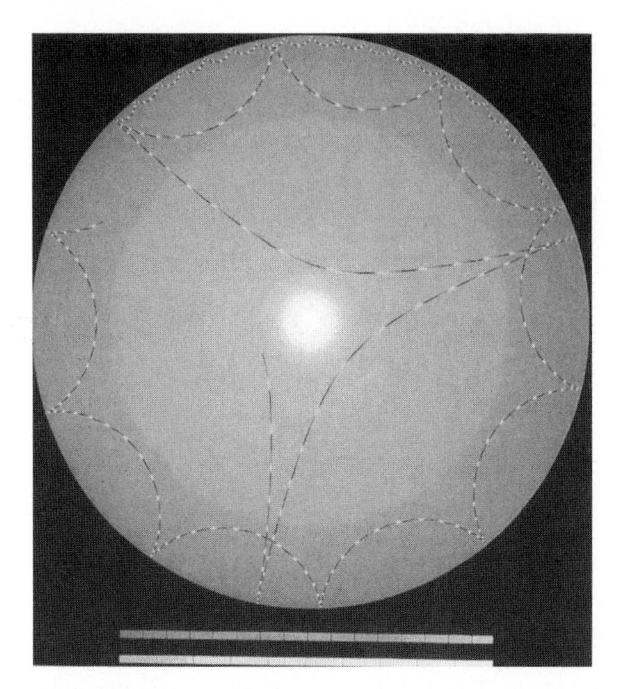

Fig. 4.14. A cutaway model of the Sun, showing refracted acoustic rays (*dashed curved lines*) and the temperature (*color, increasing from left to right along the legend bar*) and energy production (intensity) through the solar interior. Ray calculations were done by Dr. D. D'Silva. Illustration reprinted with permission from Harvey (1995, Fig. 2). Copyright 1995, American Institute of Physics (See also Plate 3)

and T increases inward (from the temperature minimum), both v_{s} and f_{λ} increase with depth.

Internal gravity waves are thought to be important only below the convection zone and are extremely weak at the surface. The *buoyant frequency*, sometimes known as the *Brunt–Väisälä frequency* for this mode of oscillations is:

$$f = \sqrt{g[(1/\gamma P)\partial P/\partial r - (1/\rho)\partial\rho/\partial r]} \tag{4.44}$$

To our knowledge, there has been no undisputed detection of g-mode solar oscillations to the present time, but Wolff (2002) suggests that maxima in a 50-year record of 10.7 cm solar radio flux data provide evidence of solar core rotation (at ~430 nHz) and g-mode beat frequencies (below ~60 nHz). The p-modes that are seen already tell a great deal about the Sun's structure; ray tracing is another informative technique, just as the arrival times of seismic waves at various locations on Earth's surface indicate the structure of the Earth's interior. A downward-propagating acoustic wave is refracted and traverses a distance over a particular time interval

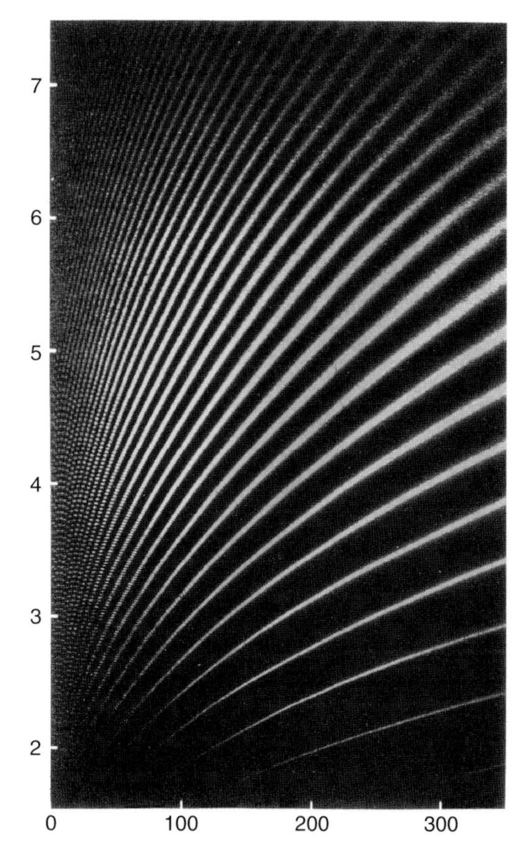

Fig. 4.15. The power spectrum of solar oscillations in the form of standing acoustic wave patterns. The frequency in mHz is plotted against the spherical harmonic degree, ℓ. In this plot, color represents depths, with *blue* indicating shallow and *red*, indicating deep, penetrating modes. Image supplied by John W. Harvey. Illustration reprinted with permission from Harvey (1995, Fig. 3). Copyright 1995, American Institute of Physics (See also Plate 4)

that depends on the conditions along the path. The time intervals and locations on the Sun's surface are determined not by discrete "quakes", as on Earth, but on correlations among the waves seen in ringed areas (annuli) around particular points. Figure 4.14 illustrates three types of refracted wave paths. As noted above, the sound speed increases inward, so ingoing (non-vertical) waves curve upward. Following the predictions of Ulrich (1970), Leibacher and Stein (1971), and of Wolff (1972), who first noted the global character of the oscillations, the first clear detection of predicted p-mode signatures was by Deubner (1975). The oscillations require the base of the convection zone, once thought to be 0.80–0.85$\Re_\odot$, to be substantially deeper (Gough 1977). The best value at present for the base of

[4.4] Determine the wavelengths of peak emission for the Planck functions of Figure 4.3, and compare the total wavelength-integrated flux emitted by these two black bodies. Finally, if the cooler of these two objects had twice the luminosity of the hotter, compute the ratio of their diameters.

[4.5] Determine the linear sizes of the sun spots, flare, and smallest features that can be discerned in Figure 4.8.

5. General Properties of the Terrestrial Planets

The terrestrial planets are confined to the inner solar system: Mercury, Venus, Earth (or Earth–Moon), and Mars. They differ from the main belt asteroids (Milone & Wilson 2008, Chapter 15.7.3) in being much more massive and from the giant planets (Chapter 12) in being much less massive; they also lack the extensive mantles of hydrogen and helium of the gas giants. The rocky, central core of Jupiter is estimated to be between 10 and 20 Earth masses, and if it became visible, might be considered the principal "terrestrial planet." However, those rocky cores are under greater pressures than are the interiors of the terrestrial planets[1], and will have different traits. For example, they are expected to be mixed with large quantities of ices, and this is true also of the smaller bodies of the outer solar system. Thus the natures of the planets of the inner solar system are sufficiently unique to be discussed as a separate group, the "inner" or "terrestrial" planets.

5.1 Overview of Terrestrial Planets

The vast bulk of the rocky material of the terrestrial planets is in the interior, so much of what we know about the material of each planet comes mainly from its bulk properties (i.e., the mean density, size, and mass) and the viewable surface. Although only the Moon and Earth have yielded sufficient material for detailed laboratory analyses, the surfaces of Mars and Venus have been examined by on-site probes and the composition has been investigated at a limited number of sites. Mercury has been seen only from fly-by missions and remains largely unknown. Its similarity to Earth's moon has made it a somewhat less well-targeted destination.

On the whole, the major planets are more spherical in shape than are the minor planets, comets, and meteoroids that permeate the larger solar system,

[1] Pressures of $\sim 10^5$ Pa (1 Pascal $= 1\,\mathrm{N/m^2} = 10^{-5}$ bar) for Earth's surface and $\sim 3.6 \times 10^{11}$ Pa for the Earth's core compare to an estimated 3.7×10^{10} for the core of Mars and 1.1×10^{12} for that of Jupiter (Zharkov and Trubitsyn 1978; Cox 2000). The physical conditions in the rocky core of Jupiter are beyond current capability to recreate them in the laboratory.

but none of them are precisely spherical. The approximate shape can be determined by direct filar micrometry (from ground-based telescopes) and, in most cases, those have been improved by direct imaging from space craft. The internal mass distributions can be and have been investigated by the perturbations on natural or artificial satellite orbits for all the planets.

The properties of planetary interiors have been investigated primarily through seismic effects (for the Earth and the Moon), magnetic field effects (for some), and the bulk properties such as radii, masses, mean densities, and surface densities (for all). In Chapter 6 and Chapter 7, we deal with the properties of the interior and how they are known; the Moon will be treated separately in Chapter 8.

Telescopic observations of planets have been made since the early seventeenth century, and still continue to provide fresh insights and discoveries. The long history of observations of the alleged planet Vulcan, of Mercury and Venus rotation rates from perceived markings, and of Martian canals from perceived linear features demonstrates clearly that observations are not foolproof, but the discoveries of the lunar maria, the gibbous phases of Venus, Galilean satellites, the moons of Mars, the planet Uranus, the first asteroid, Ceres, and Charon (the moon of Pluto) changed our perception of the solar system in lasting ways. Ground-based telescopic observations still play a role today. A "great white spot" that suddenly appeared on Saturn in the 1990s was such a discovery.

At the end of this chapter, we will touch on the visibility of planetary surfaces. We will define and discuss the significance of such terms as the *phase*, the *phase angle*, and *phase function* in studying the reflective properties of a solar system object. The surface features of the major terrestrial planets, namely Venus, Earth, and Mars, are quite similar in many respects, but there are also important differences among them, aside from the oceans of water on the Earth and the extensive atmosphere on Venus. In all the planets, impacts have played an important role in determining their physical and dynamical properties. These properties and the similarities and differences among terrestrial planets will be explored more thoroughly in Chapter 9.

Finally, the atmospheres and meteorology of the planets provide interesting differences in chemistry as well as physics. Atmospheric physics and atmospheric and ionospheric chemistry will be treated in detail in Milone & Wilson (2008, Chapters 10 and 11).

5.2 Bulk Properties

The mean densities of the planets can be obtained simply from the bulk properties of mass and radius. The mass is directly obtained by observation of the semi-major axis of a moon in the case of most of the planets or from the acceleration of a space probe in the case of Mercury, Venus, (most) asteroids, or comets. In either case, the acceleration of the small body is

$$\mathbf{a} = -\mu\hat{\mathbf{r}}/r^2 \tag{5.1}$$

where $\mu \doteq GM$, and M is the planet's mass. The planet's mean density follows:

$$< \rho >= \frac{M}{4/3\pi R^3} \tag{5.2}$$

The planet with the highest mean density is the Earth (5.515 g/cm^3 or 5515 kg/m^3), that with the lowest, Saturn (690 kg/m^3). Because different types of material have different densities (e.g., liquid water, ~1000 kg/m^3; the minerals pyroxene and olivine, ~3300; iron sulfide, 4800; and metallic iron, 7900 kg/m^3), one might suppose that determining the composition would be a simple matter of getting the correct mixture of material. The volume fraction[2] X_i of material i contributes to the overall density through the relation:

$$<\rho> = \Sigma\rho_i X_i \tag{5.3a}$$

where $<\rho>$ is the mean density[3] For instance, the planet Mercury has a mean density, $<\rho> = 5430$ kg/m^3. We can predict the fraction of metallic iron, X_{Fe} and some average "rock," of density, say, $\rho = 3500$, with a fraction $X_{\mathrm{rock}} = 1 - X_{\mathrm{Fe}}$. From (5.3):

$$< \rho >= \rho_{\mathrm{Fe}} X_{\mathrm{Fe}} + \rho_{\mathrm{rock}}(1 - X_{\mathrm{Fe}}) \tag{5.4}$$

from which one may derive[4] $X_{\mathrm{Fe}} = 0.43$, and therefore, $X_{\mathrm{rock}} = 0.57$. But this can be misleading, because the mean density describes the bulk of material that is not at standard temperature and pressure (STP) conditions. Much of it is under extremely high pressure, and associated high temperature, so that the mean density, as determined from the mass and volume, has limited value in predicting the compositional mix. Models have, however, been constructed for planetary structure which permit estimates of the *uncompressed* density. See Table 5.1, taken from Consolmagno and Schaefer (1994, Table 4.2, Ch. 4.1, p. 71),

---

[2] V_i/V_{total} where V is the volume and i a constituent.

[3] One may similarly compute the mass fraction (also referred to as X_i) from a relation of the kind,

$$\mathbf{1}/<\rho> = \Sigma\{X_i/\rho_i\} \tag{5.3b}$$

[4] With the (compressed) mean density, one can derive a mass-fraction, $X_{\mathrm{Fe}} = 0.64$ for metallic iron, assuming that to be the only form that iron takes in Mercury.

Table 5.1. Planetary densities (kg/m^3)

Planet	ρ_{com}	ρ_{uncom}	Comments
Mercury	5430	5300	Still lots of iron!
Venus	5240	4000	
Earth	5515	4100	
(*Moon*	*3360*	*3300*	*Close to the density of Earth's mantle*)
Mars	3940	3700	

Notice that the Moon has the lowest mean density among these bodies and that its compressed and uncompressed densities are about the same. Note also the mild trend of lower uncompressed mean densities with increased distance from the Sun. Recalculating the volume and mass fractions with the *uncompressed* mean density of Mercury, the (assumed) metallic iron fractions become 0.41 and 0.61, respectively.

Another observable quantity involving bulk as well as dynamic properties is the *specific angular momentum* (SAM, the angular momentum per unit mass, **h**) of a planet and its satellites, because, as in the solar system generally, the distribution of angular momentum provides important clues to the origin of the planetary and satellite systems. For the planets and for three satellite systems of the outer planets, log–log plots of SAM vs. the distance in units of the radius of the primary about which the orbiting is being done reveal that the orbiting objects fall on a straight line.

5.3 Gravitational Potential Fields

In this section we make use of conventional notation, which differs somewhat from that used in earlier chapters. If a planetary mass, M, is spherically symmetric, then a small mass, m, placed at a distance r from the centre of M has a *potential energy* (Figure 5.1)

$$U = -\frac{GMm}{r} \tag{5.5}$$

The *gravitational potential*, V, is defined by the equivalence

$$\left\{\begin{array}{l}\text{gravitational}\\ \text{potential at } r\\ \text{due to } M\end{array}\right\} \equiv \left\{\begin{array}{l}\text{potential energy per unit}\\ \text{mass for mass } m \text{ located at}\\ \text{distance } r \text{ from } M\end{array}\right\}$$

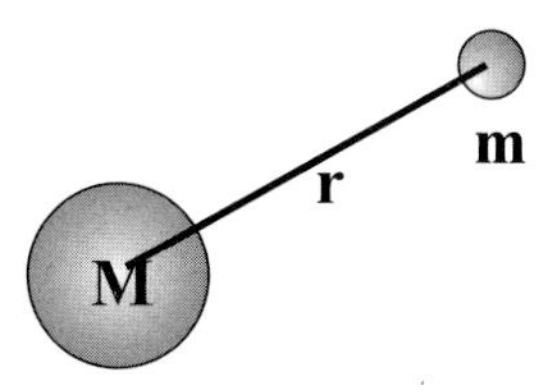

Fig. 5.1. Masses M and m have gravitational potentials and potential energies by virtue of their separation, r

That is,

$$V = \frac{U}{m} = \frac{\left[-\frac{GMm}{r}\right]}{m} = -\frac{GM}{r} \tag{5.6}$$

Then, by way of a check, the potential energy of any mass m placed in this gravitational potential is:

$$U = mV = m\left(-\frac{GM}{r}\right) = -\frac{GMm}{r},$$

which agrees with (5.5).

If the mass M is *not* spherically symmetric, the strength and direction of the gravitational acceleration, a, are influenced by the actual distribution of mass inside M and by the location of m (in terms of r, θ, and ϕ) relative to the center of M, as shown in Figure 5.2. Note the notation convention here, which differs from the usage of spherical astronomy in Chapter 2: the angle θ is called the *co-latitude* and is measured from the pole, not the equator. In the present context, the symbol ϕ is used for the longitude. V is now a function of r, θ and ϕ, and (5.6) no longer applies. It turns out that the equation for V can be written as a summation over an infinite series of spherical harmonics using *Legendre polynomials*, P_n (cos θ), and the *associated Legendre polynomials*, P_n^m (cos θ).

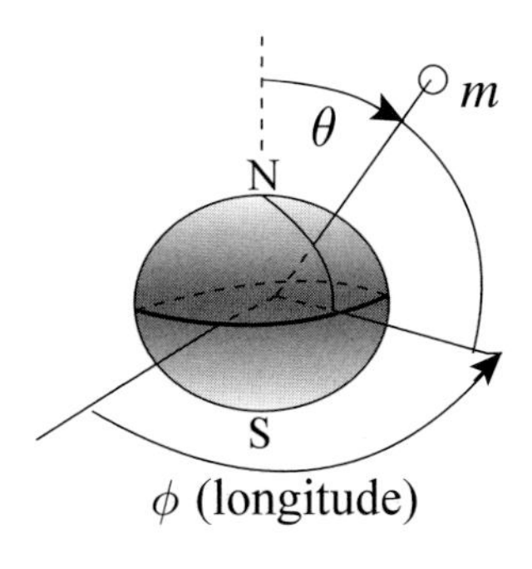

Fig. 5.2. Non-symmetric M

These are standard mathematical functions which are often encountered when using spherical coordinates. The properties of these polynomials are summarized in the addendum to this chapter (Section 5.6). With them, the potential in this spherical coordinate system can be written:

$$V = -\frac{GM}{r}\left[1 - \sum_{n=2}^{\infty}\left(\frac{a}{r}\right)^n J_n P_n(\cos\,\theta)\right] + \left[\sum_{n=2}^{\infty}\sum_{m=1}^{\infty}\left(\frac{a}{r}\right)^n (C_{nm}\cos\,m\phi + S_{nm}\sin\,m\phi)\,P_n^m(\cos\,\theta)\right] \tag{5.7}$$

where r is the distance and a is the mean radius of the planet. The polynomials are standard functions, so it follows that the coefficients must be determined by the particular mass distribution of any given planet. If we can find these coefficients from properties of the orbits of moons or of spacecraft, then (5.7) can, in principle, be "inverted" to find the mass distribution inside the planet required to produce these coefficients.

The terms in P_n do not involve the longitude, ϕ, and so describe mass distributions that are symmetric about the rotation axis but vary with latitude (as, for example, in a rotationally flattened sphere). The terms in P_n^m do involve ϕ and so describe the amount of departure from axial symmetry (longitude-dependent mass distributions).

Even-numbered J_n (J_2, J_4, ...) describes mass distributions that are symmetric about the equatorial plane (the northern hemisphere is a mirror image of the southern hemisphere), whereas odd-numbered J_n (J_1, J_3, ...) describes asymmetric distributions (such as a pear shape, for example). J_1 is missing (i.e., $n = 1$ is missing from the summation) because $n = 1$ corresponds to a bodily movement of the sphere in some direction away from the center of the coordinate system. For simplicity, we define our coordinate system to be centered on the center of the sphere, so J_1 disappears. J_2, the coefficient for the second spherical harmonic term, is called the *quadrupole moment* and measures the amount of polar flattening. It is generally the simplest to find. Figure 5.4 shows a satellite (either natural or artificial) in an inclined orbit around a planet which is rotationally flattened.

If there are no other "irregularities" in the planet, then rotation produces a mass distribution which is symmetric about the equator. The plane of the satellite's orbit "slices" the planet into two halves of equal mass, as shown in Figure 5.3, where the dotted line marks the plane of the orbit of the small body of mass m and where X_N and X_S mark the centers of mass of the northern and southern portions of the planet delineated by the plane of the orbit. When the satellite is north of the planet's equator, it is closer to the center of mass of the southern half of the planet X_S than to that of the northern half, X_N, as shown, and so feels a net off-center pull toward the south. This force can be resolved into two components,

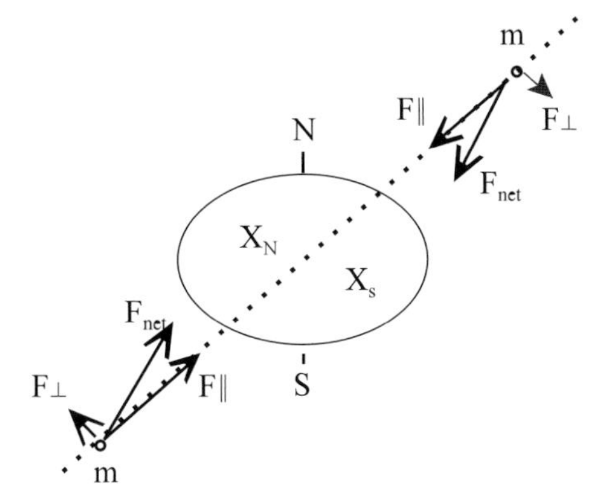

Fig. 5.3. Planet and satellite orbit plane

one in the plane of the orbit (toward the planet's center) and one at right angles to the orbit. The force toward the center is the centripetal force, the geometric description of the gravitational force of the planet that keeps the satellite in orbit, while the force at right angles in effect creates a torque on the orbit. As with a simple gyroscope spinning on one end on a desk, this gravitational torque causes the orbit to precess. The rate of precession depends on the amount by which the center of mass of each half of the planet is offset from the planet's center, and therefore the precession rate also depends on the amount of flattening and on the mass distribution inside the planet (the same amount of flattening will produce less offset in a planet with mass more concentrated toward the center).

J_2 can be calculated from the rate of precession and provides an important tool for probing the interior of a planet.

The higher-order terms and longitude-dependent terms are due to departures from the simple rotational flattening described above and give a measure of the extent to which non-hydrostatic forces (i.e., other than gravity and rotation; for example, rising mantle plumes, mountain-building due to lithospheric convergence, massive lava flows supported by stresses in the planet, etc.) have affected the mass distribution in the planet. They show up observationally as changes in the plane, orientation, size, eccentricity, and period of the satellite's orbit.

5.4 Structure of the Earth

Through the physics of wave phenomena, the structure of the interiors can be explored. The type and amount of refraction of waves are the means of exploration. Seismic events can be either natural ("quakes") or artificial, i.e., an explosive charge can be set off at some site and the response as a function of distance from that site can be recorded by seismometers set up in a net around the site. In particular, changes in wave velocity show up in a distribution of travel times with distance. The delay time as a function of position along the surface can be modeled by gradual and sharp changes of velocity in the interior and can reveal the presence of discontinuities in the refractive properties of the interior layers.

5.4.1 Seismic Studies

Seismic studies are most mature for the Earth but the Moon has seismic detectors on its surface which were placed there by the Apollo astronauts. Thus far there have been no comparable studies of Venus and Mars. Seismic events such as earthquakes release energy that displaces and distorts rock at considerable distances through basically two types of waves: *P-waves* and *S-waves*, illustrated in Figure 5.4.

P or *primary* waves are also known as pressure waves, push–pull waves, longitudinal waves, and compressional waves. They are analogous to sound waves.

S or *secondary* waves are also known as shear waves, shake waves, and transverse waves. These are analogous to light or, more generally, electromagnetic waves.

P- and S-waves are *body waves* (through the Earth); there are also *surface waves*.

When waves are generated by an earthquake, the P-waves travel faster than the S-waves and arrive first at any given location. The speed of a wave depends on:

1. The type of wave (P or S)
2. The density of the medium

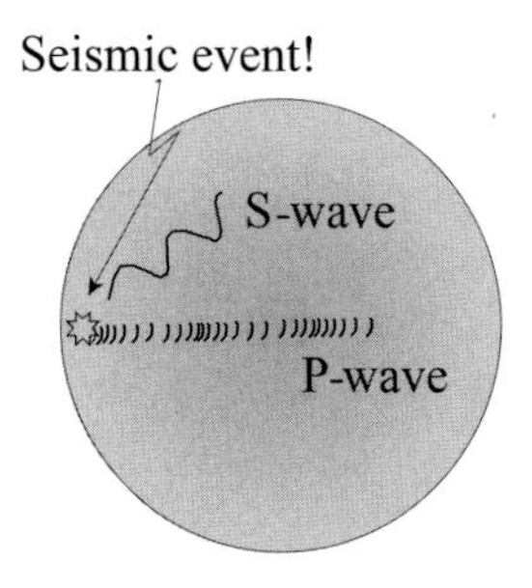

Fig. 5.4. P- and S-waves

3. The state of the material (solid, liquid, or gas)
4. The composition (granite, basalt, iron, ...)
5. The mineral phase (e.g., the wave speed is different in graphite and diamond, even though they are both forms of carbon)
6. The compressibility, K, for P-waves, and the rigidity, μ, for P- and S-waves. K and μ depend on the density, and the density, in turn, depends on the composition and the mineral phase (as well as the temperature and the pressure), so in fact the wave speed depends very strongly on the density

Because pressure ("P") and shear ("S") waves differ in their speeds through material (e.g., the S-waves do not propagate through fluids), the physical state of the material can be found. Thus the Earth has an outer core that is liquid, revealed by a "shadow zone" in which the S-waves are not seen on the surface. The structure of the interior is then deduced through the solution of equations describing the expected physics of the interior: an equation of state, which relates the density, pressure, mean molecular weight, and temperature; an equilibrium equation that relates the pressure with the weight per unit area; expressions of the conservation of mass; and an expression which describes heat flow.

The travel paths of seismic waves are decided by the refraction properties of the medium. Snell's law applies:

$$n_1 \sin\theta_1 = n_2 \sin\theta_2 \tag{5.8}$$

where the index of refraction, n, depends on the speed of the wave.

As Figure 5.5 demonstrates, a wave is bent *toward* the normal if it travels from a region of lower refractive index to a region of higher refractive index (i.e., higher wave speed to lower wave speed), because $n_2 > n_1 \Rightarrow \theta_2 < \theta_1$. It is bent *away from* the normal if it travels from a region of higher refractive index to a region of lower refractive index (corresponding to a region with lower wave speed to one of higher wave speed), because

$$n_1 > n_2 \Rightarrow \theta_2 > \theta_1 .$$

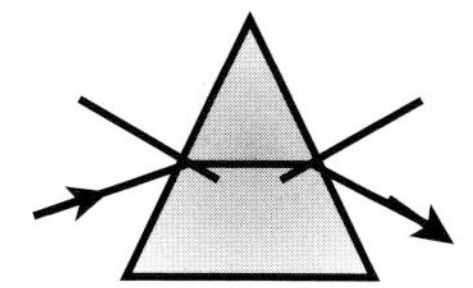

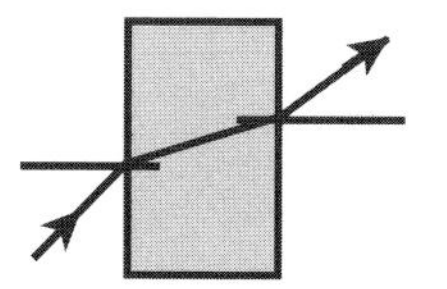

Fig. 5.5. Snell's law

The path in a layered sphere involves additional considerations. Density generally increases with depth into the Earth so, except at certain discontinuities, the wave speed, v, increases with depth. The refractive index thus generally decreases with depth. We consider five conditions:

1. A layered sphere, v constant in each layer (Figure 5.6): the ray refracts *away* from the normal going in and *toward* the normal going out.
2. A sphere in which v increases smoothly with depth (Figure 5.7): this is the same as if the layers were infinitesimally thin. The refraction then becomes continuous.
3. A seismic discontinuity at which v suddenly decreases and afterward continues to increase (Figure 5.8): the refractive index suddenly increases, so the wave refracts toward the normal.

   Also, both reflection and refraction take place at any boundary.
4. Liquid layer: S-waves cannot propagate unless there is a restoring force in the transverse direction. No such force exists in a liquid (a liquid has no "rigidity"), so S-waves cannot propagate in a liquid. No S-waves are observed to travel through the outer core, so the outer core is known to be liquid.

   It appears to be molten iron (Figure 5.9).
5. Shadow zone (see Figures 5.10 and 5.11): The speed of P-waves in the liquid outer core of the Earth is much less than in the mantle just above it (13 km/s in the mantle, 8 km/s in the core); v thus decreases suddenly at the core–mantle boundary.

As shown in Figures 5.10 and 5.11, refraction at the core–mantle boundary creates a shadow zone where no seismic waves will be received. (Body waves may reach there by other processes, such as reflection from the inner core, but these are much weaker than the "direct" P-waves so the shadow zone is still easily observable.) The speed of the wave determines its behavior, so the time intervals for the wave to reach various locations from the site of origin provide the data for modeling the refraction of the layers through which it passes.

If there is a seismic discontinuity in which v increases suddenly at some depth, then a station located within a certain range of distances from an earthquake can receive waves by three different travel paths (Figure 5.12):

Speed increases smoothly with depth until the discontinuity is reached, so the station receives waves as follows.

A. Seismic wave takes the most direct path. This has the slowest average v but the shortest path; it arrives first.

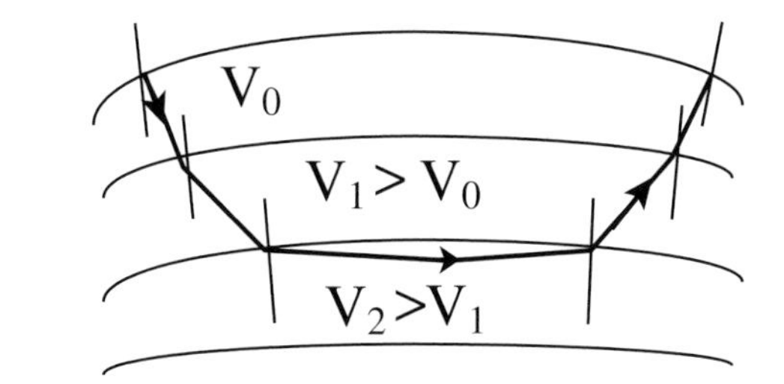

Fig. 5.6. Constant v in each layer

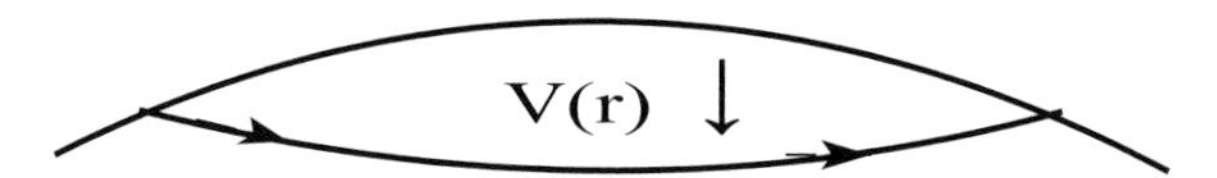

Fig. 5.7. v increasing smoothly with depth

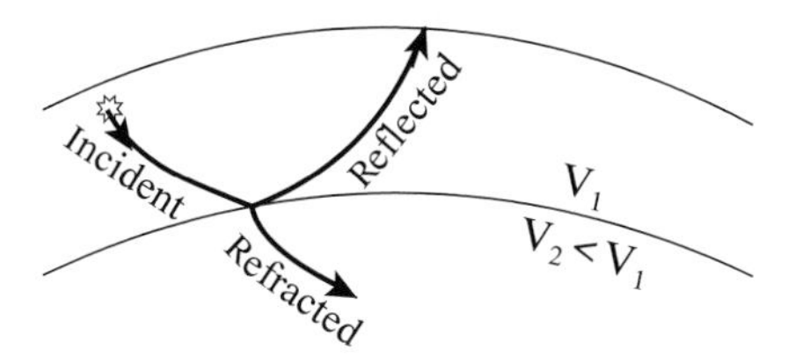

Fig. 5.8. v increase discontinuous

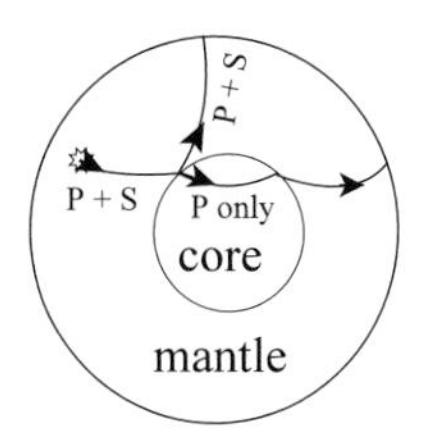

Fig. 5.9. Liquid layer passes only P-waves

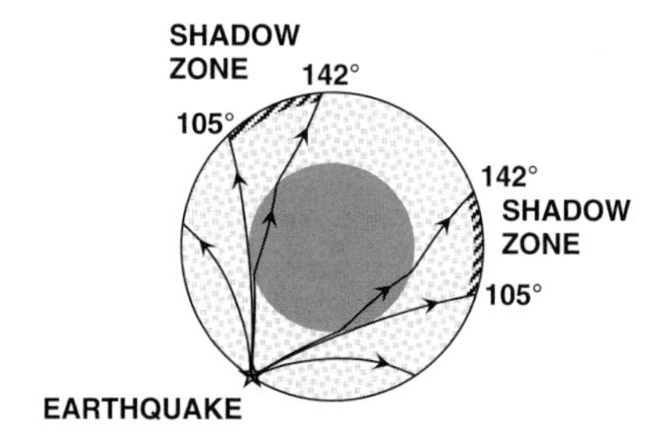

Fig. 5.10. $v_{\text{liq core}} << v_{\text{mantle}}$. Waves that just miss the core boundary return to the surface 105° from the earthquake, whereas those that just contact the core boundary refract into the core and out again, and return to the surface 142° from the earthquake. Between the two is a shadow zone in which no direct P waves are seen

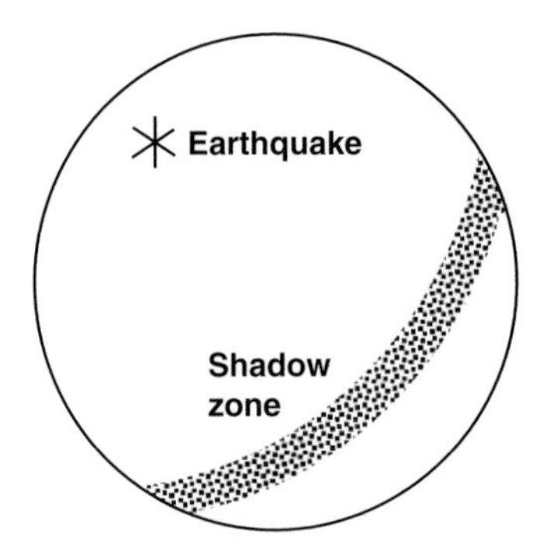

Fig. 5.11. No seismic waves in shadow zone

B. Seismic wave reflects from the discontinuity. It has faster average speed (travels deeper) but along a longer path; it arrives later than A.

C. The wave refracts into the faster layer, then out again. It has a longer path than B, but the higher speed in the lower layer more than compensates for the longer path; it arrives before B, but after A. The arrival times at station P are plotted in Figure 5.13.

If we move P closer to the earthquake site, then the travel path of wave C in the lower layer decreases, eventually becoming zero at point P_1 in Figure 5.14. The travel times for waves B and C thus approach each other, becoming equal at P_1.

If we move P further from the earthquake, then the lengths and travel times for paths A and B approach each other, becoming equal at point P_2 in Figure 5.14.

If we plot the arrival times of waves observed at various stations as a function of the distance of each station from the epicenter of the earthquake (the point on the surface of the Earth directly above the earthquake), then we find the arrival time plot as shown in Figure 5.15

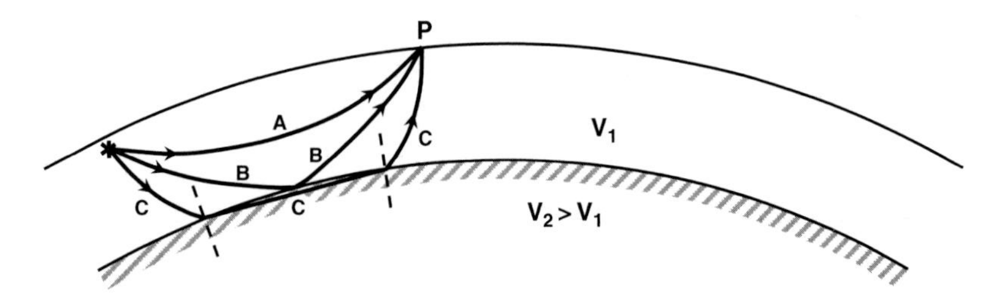

Fig. 5.12. Three wave paths from earthquake site to Station P

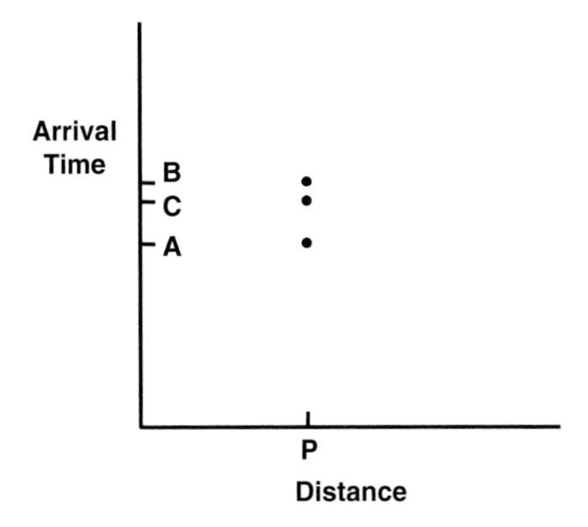

Fig. 5.13. Arrival times of seismic waves at station P

Plots such as these can be used to find wave speed as a function of depth into the Earth.

5.4.2 The Adams–Williamson Equation

A derivative expression, the Adams–Williamson equation, relates the velocities of the P and S-waves deduced from the seismic models to the density gradient, $\mathrm{d}\rho/\mathrm{d}r$, at a distance r from the center. The usual procedure is to consider the interior to be a series of shells of thickness $\mathrm{d}r$ and to express the change in density, $\mathrm{d}\rho$, across that shell. The density depends also on the molecular weight of the material in the interior, and this is not known exactly. However, it can be guessed. The equation is derived as follows:

Start with the equation of hydrostatic equilibrium:

$$\frac{\mathrm{d}P}{\mathrm{d}r} = -\rho g = -\frac{GM(r)}{r^2}\rho \tag{5.9}$$

where $M(r)$ is the mass enclosed within radius r.

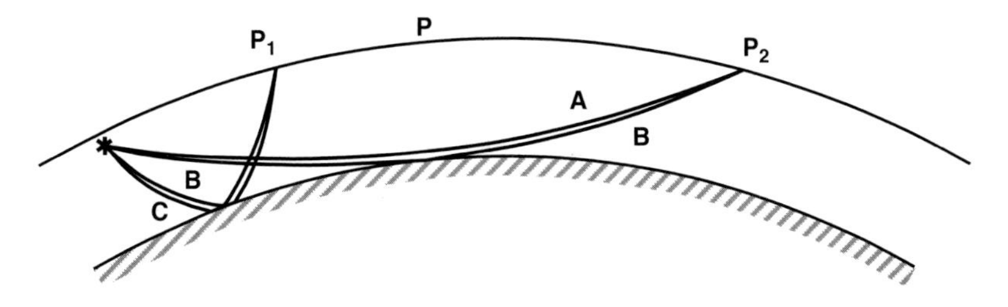

Fig. 5.14. Moving station P change the travel times

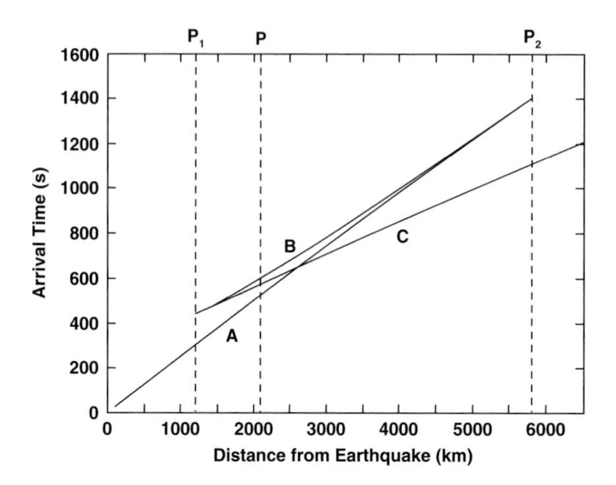

Fig. 5.15. Arrival time of S-waves as a function of the distance of the observing station from the earthquake, measured along the surface of the Earth. The data are from a simplified two-layer model of the outer mantle with a seismic discontinuity 670 km below the surface, and S-wave speeds 4.0 km/s above the discontinuity and 6.0 km/s below

Define the *bulk modulus*, k, through the relation

$$\frac{\mathrm{d}\rho}{\mathrm{d}P} = \frac{\rho}{k} \tag{5.10}$$

Rewriting (5.10), substituting from (5.9), and dividing top and bottom of the RHS by ρ, we get:

$$\frac{\mathrm{d}\rho}{\mathrm{d}r} = \frac{\mathrm{d}\rho}{\mathrm{d}P}\frac{\mathrm{d}P}{\mathrm{d}r} = \frac{\rho}{k}\frac{\mathrm{d}P}{\mathrm{d}r} = -\frac{GM(r)\rho}{\frac{k}{\rho}r^2} \tag{5.11}$$

Then, making use of the relations between these quantities and the primary and secondary wave speeds,

$$\nu_p{}^2 = \frac{m}{\rho} = \frac{k + \frac{4}{3}\mu}{\rho}, \quad \nu_s{}^2 = \frac{\mu}{\rho} \tag{5.12}$$

where m is the *elastic modulus* and μ is called the *rigidity* or *shear modulus*,[5] so that, combining the (5.12) expressions,

$$\nu_\mathrm{p}{}^2 = \frac{k}{\rho} + \frac{4}{3}\nu_\mathrm{s}{}^2 \tag{5.13}$$

[5] For the derivation of these quantities in terms of stresses and Young's modulus, see, for example, Stacey (1969), pp. 85–86.

and, rearranging (5.13),

$$\frac{k}{\rho} = {\nu_\mathrm{p}}^2 - \frac{4}{3}{\nu_\mathrm{s}}^2 \tag{5.14a}$$

we arrive at what has been called the “fundamental equation of geology,” the *Adams–Williamson* equation:

$$\frac{\mathrm{d}\rho}{\mathrm{d}r} = -\frac{GM(r)\rho}{r^2\left({\nu_p}^2 - \frac{4}{3}{\nu_\mathrm{s}}^2\right)} \tag{5.15}$$

Now we go on to discuss how to solve this equation in order to find the density, ρ, as a function of r inside the Earth.

Equation (5.15) tells us the change ($\mathrm{d}\rho$) in ρ over some small distance $\mathrm{d}r$:

$$\frac{\mathrm{d}\rho}{\mathrm{d}r} = f \quad \Rightarrow \quad \mathrm{d}\rho = f\,\mathrm{d}r \tag{5.16}$$

Therefore, to find ρ as a function of r, we need to integrate (5.16).

$$\rho = \int \mathrm{d}\rho = \int f\ \mathrm{d}r. \tag{5.17}$$

This involves numerical integration; we describe the procedure in a step-wise manner:

First, divide the Earth into many thin shells, as shown in Figure 5.16.
Each shell has uniform composition and density.
Then the density, ρ, in each shell depends on:

1. mineral composition;
2. compression of the shell by the weight of all the shells above it.

Decide on a model for the Earth’s composition as a function of r, for example: olivine/pyroxene mantle, iron core.

Note that we can express the total mass, M_r, interior to any radius, r, either by:

$$M_r = \int_{a=0}^{r} \mathrm{dm} = \int_{a=0}^{r} \rho(a)\ \mathrm{d}V = \int_{a=0}^{r} \rho(a) 4\pi a^2\ \mathrm{d}a \tag{5.18}$$

or by

$$M_r = M_\mathrm{E} - \int_{a=r}^{R_\mathrm{E}} \mathrm{d}m = M_\mathrm{E} - \int_{a=r}^{R_\mathrm{E}} \rho(a) 4\pi a^2\ \mathrm{d}a \tag{5.19}$$

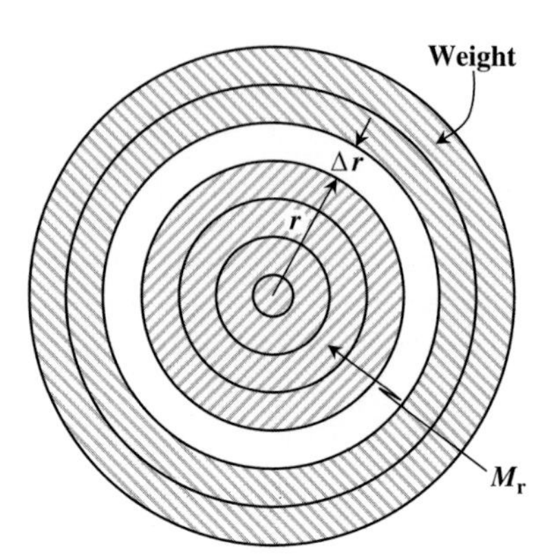

Fig. 5.16. Shells for numerical integration of (5.17)

where M_E and R_E are the Earth's mass and radius respectively. Equation (5.19) is more useful than (5.18), because it involves only quantities above the radius r, and these are known at any given r, as described below.

As we are going to integrate numerically, we need numbers, and an infinitesimal quantity like da or dr is not a number. Therefore, we need to convert (5.15) and (5.19) to *finite difference equations.*

Taking (5.19) first, we replace da by the width, Δr_i, of the ith shell:

$$M_r = M_{\mathrm{E}} - \sum \rho_i 4\pi {r_i}^2 \, \Delta r_i \tag{5.20}$$

We know M_{E} and R_{E}, and we know ρ at the surface of the Earth. Now we can go through an iterative loop, working inwards from the Earth's surface.

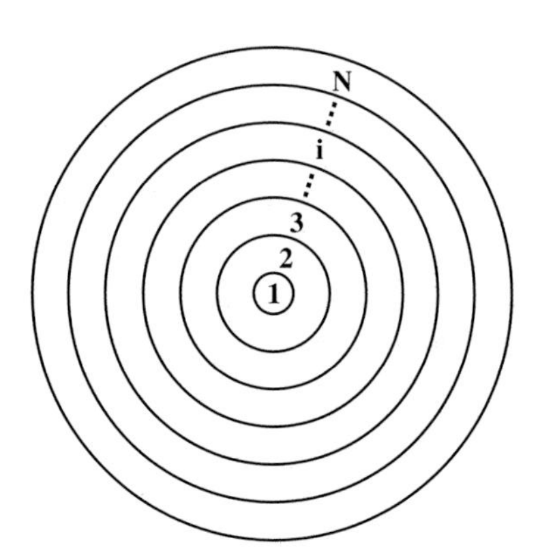

Fig. 5.17. Setting up N Shells

LOOP:

1. Decide on a thickness Δr_i for this shell.
2. Use (5.20) to calculate M_r (Note: the RHS is known). This involves numerical integration.
3. Rewrite the Adams–Williamson equation as a finite difference equation:

$$\Delta\rho_i = \left\{ \frac{GM_r\rho_i}{{r_i}^2\left[(\nu_P)_i{}^2 - \frac{4}{3}(\nu_S)_i{}^2\right]} \right\} \Delta r_i \tag{5.21}$$

where M_r was found above, and v_P and v_S are known from the seismic data.

This gives $\Delta\rho_i$, the change in ρ across the shell, so ρ for the next shell inward is:

$$\rho_{i-1} = \rho_i + \Delta\rho_\mathrm{i}. \tag{5.22}$$

We continue looping until we either run out of mass ($M_r = 0$) or run out of radius ($r_i = 0$). Because M_r has to be 0 exactly at $r = 0$, we have to adjust the composition and other input assumptions until they both go to zero at the same time. We now have the density, ρ, as a function of r. The model also has to allow for mineral phase changes, e.g., the crystal structure of olivine changes at about 150–200 kbar pressure to a new crystal structure; it is then called *spinel* and is about 10% denser than the original crystal structure (*olivine*) at the same pressure. See Chapter 7 for a discussion of mineral structure.

5.4.3 Moments of Inertia

The moment of inertia I, of the Earth depends on the distribution of mass (and therefore density) inside the Earth:

$$I = \sum m_i\,{r_i}^2 \tag{5.23}$$

For a uniform sphere of radius R and mass M, one may show that

$$I = (2/5)\;MR^2 \tag{5.24}$$

In general, however,

$$I = KMR^2 \tag{5.25}$$

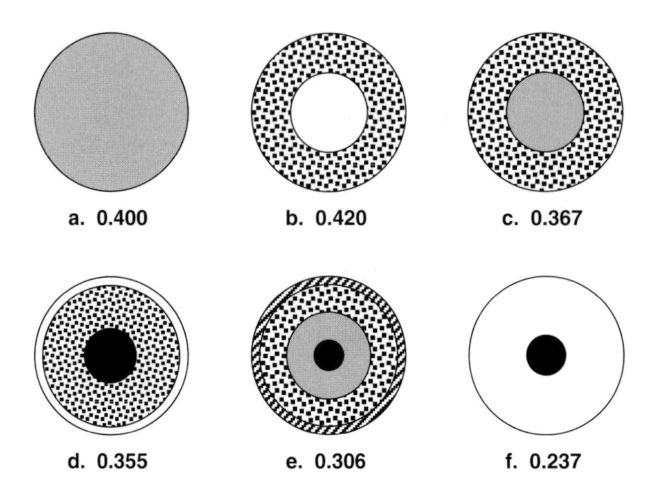

Fig. 5.18. The effect of density distribution on K. Outer radius, 1 unit. **a**. *Undifferentiated asteroid or comet.* Uniform density. **b**. *Density inversion.* Core: $r = 0.5$, $\rho = 0.5$; mantle: $\rho = 1.00$. **c**. Core: $r = 0.5$, $\rho = 2.00$; mantle: $\rho = 1.00$. **d**. *Callisto model.* Core: $r = 0.34$, $\rho = 3.5$; mantle: $r = 0.89$, $\rho = 2.1$; crust: $= 1.0$. **e**. *Terrestrial planet model.* Inner core: $r = 0.19$, $\rho = 12.8$; outer core: $r = 0.55$, $\rho = 11.2$; lower mantle: r $= 0.895$, $\rho = 4.9$; upper mantle: $\rho = 3.6$. **f**. *Jovian planet model.* Core: $r = 0.25$, $\rho = 20$; mantle: $\rho = 0.4$. (Densities in g/cm^3.)

See Figure 5.18 for the effect of different density distributions on K. Note that I is lower if most of the mass is close to the center (in the core) and I is higher if most of the mass is further out (in the mantle). We can calculate the moment of inertia for our model of density as a function of radius and compare that to the observed value. If the two are different then we can adjust our model to bring it into agreement with the observations. With regard to orbiting satellite observations, we note that the second harmonic can be written in terms of the moments of inertia:

$$J_2 = [I_z - I_{x,y}]/(M\, r^2) \tag{5.26}$$

where I_z is the moment of inertia about the rotation axis of the planet and $I_{x,\,y}$ is the moment of inertia about an axis in the equatorial plane. Some of the effects on the orbit (see Chapter 3 for definitions) are as follows:

$$\mathrm{d}\omega/\mathrm{d}t = -(3/2) J_2 (n/p^2)[(5/2)\,\sin^3\, i - 2] \tag{5.27}$$

where $p \equiv a(1 - e^2)$,

$$\mathrm{d}\Omega/\mathrm{d}t = -(3/2)\, J_2 (n/p^2)\, \cos\, i \tag{5.28}$$

and on the mean motion, and thus the position of the object in its orbit,

$$[t - T_0](\mathrm{d}n/\mathrm{d}t) = -(3/2) J_2 (n/p^2)[(5/2)\,\sin^3\, i - 2] \tag{5.29}$$

5.4.4 Models of the Interior

The integration of the Adams–Williamson equation is used to produce a march of the density with distance from the center. The models of the planet are constrained by the surface boundary layer where the density, pressure, and temperature drop sharply, and by the bulk properties of the planet: the mean density and radius. Samples of models may be found at the end of this module. By means of equations of state, which relate the temperature and pressure to the density, T and P can be similarly obtained for the interior as functions of r.

The densities are seen to increase through the interior, along with the temperature and the pressure. The behavior of various materials under the conditions that prevail in the interior has led to much discussion about the composition at various depths. One important question is why different materials *should* be located at different depths. Some bodies in the solar system seem to be of uniform composition. Such bodies are said to be *undifferentiated.*

A process known as *differentiation* is believed to have taken place early in planetary history. This process is the separation out of denser material from less dense material when the body was in an originally molten, or at least highly plastic, state. The seismic modeling confirms this view of the Earth's interior, as does our knowledge of meteorites, some of which show evidence of such differentiation also (see Chapter 15).

The Earth's core (discussed in detail below) is expected to contain primarily iron with substantive amounts of nickel, mixed with sulfur, oxygen, and other elements. The mantle and crust of the Earth, on the other hand, are much richer in silicates. With an assumption about the density of the central core, the differences in density can be added, shell by shell, to determine the interior density at every radius from the center. In this way, the interior structure can be induced. Thus the core takes up the inner 3500 km from the center (the outer 2/3 of which is assumed to be liquid), the mantle approximately 2900 km, and the crust a mere 50–100 km. The mantle itself has structure: in the upper mantle, a deeper, more fluid component (the *asthenosphere*) underlies the *lithosphere*, which includes the crust.

The lithosphere "rides" on the asthenosphere in the form of "plates" driven by convective regions in the asthenosphere. The effect is to produce regions of *orogenesis*, or mountain-building, and *subduction*, regions where one plate (e.g., an oceanic plate) flows below another (typically a continental plate). The result is continental drift, in which continents move at rates of the order of 1 cm/y, and where areas which are subducted are replaced by new influx of material along mid-oceanic rises, which produces seafloor spreading. The vent areas are outgassing, warm, and rich in suphur and other material, and it is suspected that life began on the Earth in just such areas; in any case, there is a variety of living organisms currently found at such sites. Lava which spreads out away from the upwelling areas rapidly falls below the Curie point, and the ferro-magnetic material in it aligns with the local magnetic field in effect at the time and remains that way. Thus, the magnetic reversals in the Earth's field can be studied in the remnant magnetism trapped in successive ocean floor spread lava.

According to evidence from geomagnetism, seismology, high-pressure experimentation, and cosmochemistry, the Earth's core is composed of almost pure iron.

The major points are as follows:

1. According to many models, the density of the core is 10–13 g/cm^3 (10,000–13,000 kg/m^3). Silicates cannot be compressed to this density by the pressures expected in the Earth's core.
2. The fluidity and electrical conductivity of the silicates in the Earth's mantle are too low to produce the observed magnetic field of the Earth. The magnetic field is therefore produced by the core. This requires the core to be both liquid and metallic.
3. The seismologically observed density and P-wave velocity (sound speed) in the core are close to those of iron measured in the laboratory at similar temperatures and pressures. (These temperatures and pressures can be produced in the lab in a special device called a diamond-anvil cell.)
4. Cosmically, iron is by far the most abundant element having these properties.

5.4.4.1 The Size of the Earth's Core The fraction of the Earth taken up by the core can be summarized as follows:

	Outer core (%)	Inner core (%)
By volume	15.7	0.7
By mass	30.8	1.7
By number of atoms	15.0	0.8

5.4.4.2 The Molten Outer Core The state of this core must be molten. The evidence includes:

1. The existence of the Earth's magnetic field, as described above.
2. S-waves do not propagate through the outer core, and are not seen.
3. Results from analysis of free oscillations of the Earth as a whole.
4. The character of the Earth's nutation (small wobbles in the direction of the Earth's spin axis): if you spin an egg, it behaves differently depending on whether it is hard-boiled (solid) or raw (fluid inside the shell); similarly, the spinning Earth responds differently to gravitational tugs depending on whether it is entirely solid or has a large molten core.

5. Convective and cyclonic motions occur in the fluid outer core, producing the Earth's magnetic field through dynamo action. Speeds appear to be about 10 km/y.

5.4.4.3 Non-iron Composition of the Core The observed density of the core is $10\pm2\%$ less than that expected for pure iron at the core's temperature and pressure. At least one lighter element (possibly more) is therefore mixed in as an impurity, probably forming an alloy. The element or elements involved are not known. Those that have been suggested (in the form of pro and con arguments) are:

(1) Sulfur (S)

For:

- S alloys easily with iron, forming sulphides, at both low and high pressures.
- S lowers the melting temperature of the iron alloy compared to pure iron. This favors the formation of the observed solid inner core because, as the core temperature decreases over geological time, pure iron can separate out, solidify, and sink, forming the inner core, while the alloy in the outer core remains molten.
- Iron sulphides are found in meteorites ($0.7\pm0.5\%$ by weight of the iron).
- S is depleted in the mantle compared to cosmic abundances. Enhanced sulfur in the core would account for part of this depletion.

Against:

- S in the core cannot explain all of the sulfur depletion in the mantle, even if sulfur makes up the entire light-element content of the core. At least some sulfur therefore must have been lost to space as a volatile during the formation of the Earth. (The case for sulfur in the core would be strengthened if none could have been lost to space, because then the only place the "missing" mantle sulfur could have gone would be to the core.)
- If sulfur makes up a significant fraction of the light-element content of the core, then it would in fact be more abundant in the Earth as a whole than some elements which are less volatile (less easily lost to space) than sulfur.

(2) Oxygen (O)

For:

- O is definitely present, because chemical reactions at the core–mantle boundary necessarily add oxygen to the core. This oxygen will be mixed through the core by convection. (The question is, is it present in significant amounts?)

- O is abundant in the Earth (58±2% of the mantle by number of atoms).
- FeO becomes metallic at the high pressures found in the Earth's core, and therefore alloys easily with the molten iron there.

Against:

- O does not easily alloy with iron at low pressures. It could therefore not have alloyed with iron until well after the core had begun to form, and perhaps not until after the accretion of the Earth was complete. This may leave chemical reactions at the core boundary as the major source of oxygen in the core, and this may not be significant.

(3) Silicon (Si) and Magnesium (Mg)

For:

- Necessarily added by chemical reactions at the core–mantle boundary, as with oxygen.
- Si and Mg relatively abundant (18±7% of the mantle by number of atoms).

Against:

- Smaller amounts available compared to oxygen.
- Si and Mg appear to be less reactive with molten iron than oxygen.

(4) Hydrogen (H):

For:

- H is abundant in the universe, and also possibly at the Earth's surface during accretion and core formation.
- H alloys with iron easily at high pressures.

Against:

- Most of the Earth's hydrogen was lost to space; the amount (if any) in the core cannot explain the observed deficit compared to cosmic.

- The presence of H in the core seems to require the Earth to remain cold during accretion and subsequent core formation, to prevent the hydrogen from being lost to space. This is contrary to expectation.

(5) Nickel (Ni)

For:

- Ni is present in iron meteorites (8±7% by weight).
- Ni is depleted in the mantle by an amount which would be completely accounted for if the core contained about 4% by weight of nickel.
- Ni alloys easily with iron at both low and high pressures.
- Ni lowers the melting point of the iron alloy compared to pure iron and so favors formation of a solid inner core (as noted above for sulfur).

Against:

- Ni has almost the same density as iron, so it does not affect the need for light elements discussed above. In fact, there is *no* geophysical observation that directly indicates the presence of nickel in the core (or any that excludes it).

Clearly there are many candidates, some more likely than others.

5.4.4.4 The Inner Core The inner core is apparently solid. There are two main lines of evidence:

1. S-waves appear to propagate through the inner core; this is not possible in a liquid.
2. P-wave velocities in the inner core are systematically higher parallel to the Earth's rotational axis (the polar or N–S direction) than parallel to the plane of the equator (Calvet et al. 2006 and references therein). A similar anisotropy is found in the uppermost part of the mantle and is caused by olivine crystals having a preferred direction of orientation (produced by shear associated with convection and plate tectonics). The same mechanism could operate in the inner core if it is also tectonically active (solid but with plastic flow), because solid iron is expected to have a hexagonally close-packed (hcp) crystal structure at the inner core's temperature and pressure, and hcp iron is known to have anisotropic elastic properties.

At first sight, the existence of a solid inner core seems to contradict the fact that temperature cannot decrease inward in a solid or liquid, non-degenerate planet. This apparent discrepancy is resolved by the fact that the melting temperature, T_M, of iron and its alloys increases with increasing pressure, as illustrated schematically in Figure 5.19. The actual temperature, T_r, in the core increases more slowly with decreasing radius than does T_M, with the result that $T_r < T_M$ in the inner core and the inner core is solid.

Because of the high thermal conductivity of solid iron, the inner core is expected to approach isothermality on a timescale of the order of 10^9 years (Williams et al. 1987 and references therein). An age for the inner core can be derived from the fact that it is believed to play a significant role in the production of the Earth's magnetic field, and paleomagnetic measurements show a strong terrestrial magnetic field for at least the past 2.5–3.5 billion years. Although this may be insufficient time to achieve full isothermality, measured data suggest a maximum temperature difference between the center and outer boundary of 300 K or less, or about 10% or less of the temperature difference across the outer core (Figure 5.19).

5.4.4.5 Formation of the Inner Core Impurities usually lower the melting point of pure iron. An example is shown in Figure 5.20, where the melting temperature (T_M) at 1 atm pressure decreases with increasing percentage of sulfur in iron, by weight. T_M is much higher at 3 Mbar pressure, but the behavior is similar.

The melting temperature, T_M, of the core material as a function of pressure (and therefore of radius in the Earth) provides a lower limit to the geotherm, T_r, in the molten, outer core and an upper limit in the solid, inner core, and equals T_r at the inner core–outer core boundary (IOB). Thus, T_M provides an important constraint on the geotherm. However, experimental and theoretical uncertainties make T_M uncertain, as do uncertainties in the composition of the core material (Section 5.4.4.3); see Boehler (2000) for a review.

Static measurements of T_M in laser-heated diamond-anvil cells have been conducted for pure Fe up to pressures as high as 2 Mbar and in shock-heated

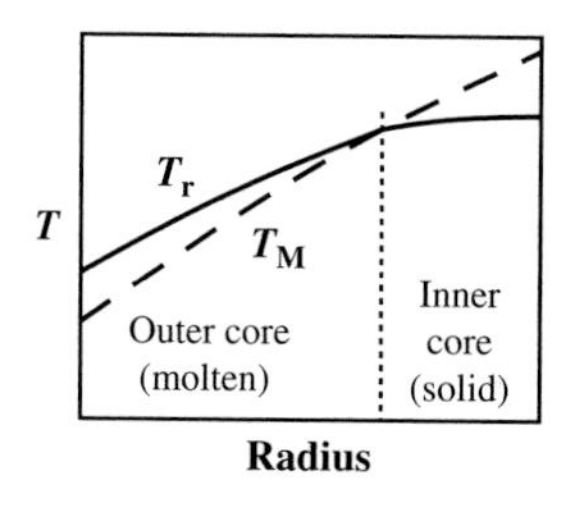

Fig. 5.19. Actual temperature, T_r (*solid line*), as a function of radius in the Earth's core compared to the melting temperature, T_M (*dashed line*)

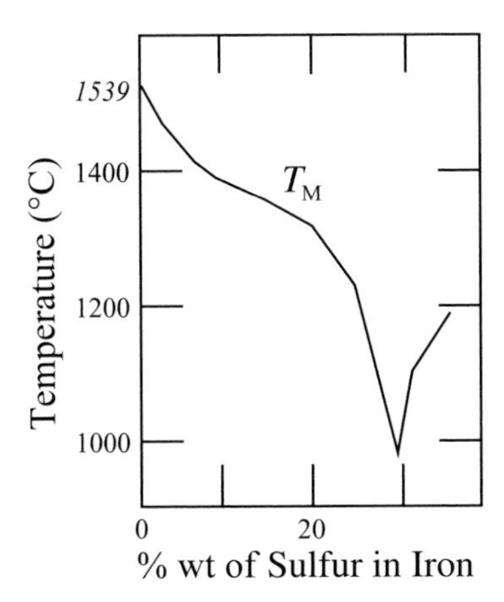

Fig. 5.20. The percentage of sulfur by weight vs the melting point of the iron mixture at 1 atm pressure

diamond-anvil cells up to about 4 Mbar, and to lower pressures for Fe–FeS systems including Fe_2S and Fe_3S, and for Fe–FeO systems. For comparison, the pressure is about 1.36 Mbar at the core–mantle boundary (CMB) and 3.30 Mbar at the IOB. Shock heating measurements are generally considered less accurate than static measurements because of various uncertainties, including the optical and thermal behaviors of the window material during the shock (Boehler 2000), but provide important constraints on the melting curves.

Several static diamond-anvil studies (Boehler 2000 and references therein) agree on T_M for pure iron in the range 3100–3400 K at the CMB, and, by extrapolation to higher pressures, 4200–5000 K at the IOB (Fig. 5 of Boehler 2000). As noted above, the presence of S or both S and O can lower T_M by a few hundred K. (Earlier studies indicated that FeO had a higher T_M than pure Fe, but more recent results suggest that, at core pressures, T_M is essentially the same for both FeO and Fe.) With T_M below 5000 K at the IOB, and assuming temperature changes adiabatically throughout the outer core, T_r on the core side of the CMB is below 4000 K.

The inner core grows in size with time as material from the outer core solidifies onto it. When solidification is slow, lighter-density components tend to be excluded from the solid phase, so the core is denser and purer iron than the outer core, and the fraction of lighter materials in the outer core slowly increases with time.

The P- and S-wave speeds and the density as a function of radius are shown for the mantle in Figure 5.21 and for the entire planet in Figure 5.22. Finally, the march of pressure, gravitational acceleration, and density through the Earth for the preliminary reference model are plotted in Figure 5.23.

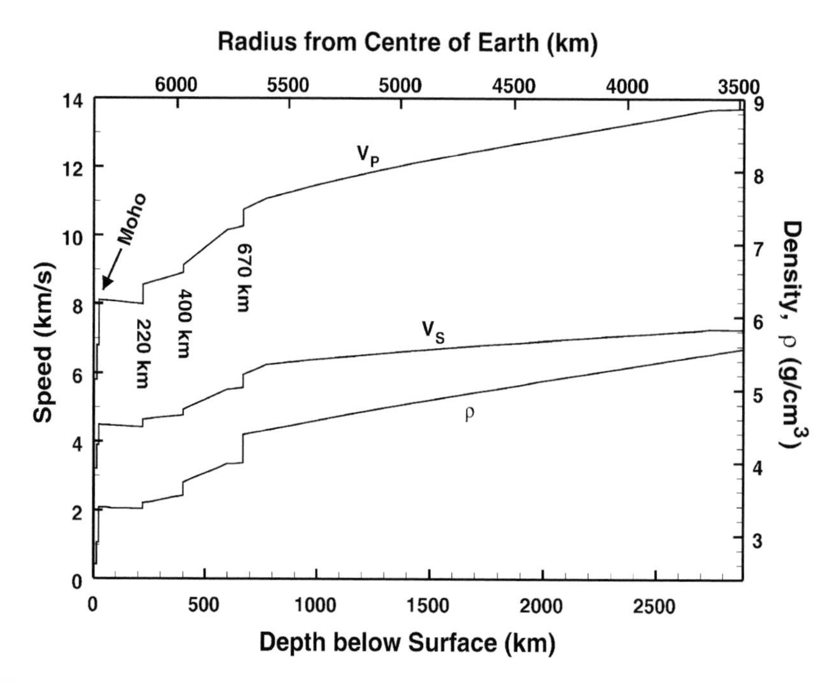

Fig. 5.21. The variation of P- and S-waves and the density through the crust and the mantle. Note the trend for all three quantities to increase with depth. Produced from data published by Dziewonski and Anderson (1981, Table II)

Some recent analyses of deep-Earth seismic data (Ishii and Dziewonski 2002, 2003; Beghein and Trampert 2003; Calvet et al. 2006) suggest that the anisotropy of the inner core displays a marked change in character in its innermost part ($r < 300$–$450\,\mathrm{km}$). Although several different studies agree on the existence of this innermost inner core, its character remains uncertain. Ishii and Dziewonski (2003), from inverting travel-time data for P-waves and using a ray approximation, find that the fast axis in the innermost inner core could be tilted as much as 55° from the Earth's rotation axis. Even assuming that the fast axis is parallel to the Earth's rotation axis, as it is in the outer part of the inner core, they still find a distinct innermost inner core in that the anisotropy is noticeably stronger. Whether the anisotropy change occurs sharply at a boundary or gradually over a range of radii is unclear. Beghein and Trampert (2003) find the fast axis in the Earth's equatorial plane for P-waves and parallel to the Earth's rotation axis for S-waves. Calvet et al. (2006) perform an extensive analysis of P-wave data using a more thorough method than previous investigators. Their results agree with previous models for the outer part of the inner core, but for the innermost inner core the data are consistent with any of three basic models: (1) a weak anisotropy with the fast axis in the equatorial

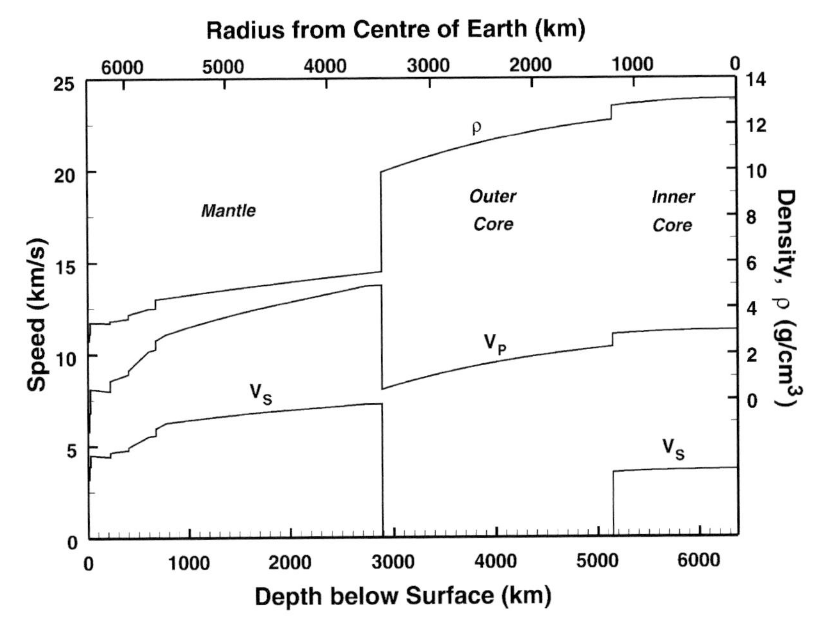

Fig. 5.22. As per Figure 5.21 but for the entire planet. Note the sudden decrease in both P and S velocities at the boundary between the core and the mantle, and the absence of S-waves in the outer core. Produced from data in Dziewonski and Anderson (1981, Table II)

plane; (2) near isotropy; and (3) a strong anisotropy with the fast axis parallel to the Earth's rotation axis. In all three cases, the innermost inner core appears distinct in its properties from the outer part of the inner core, despite the lack of constraint on these properties.

This innermost inner core (IMIC) could arise through one of at least three scenarios (Ishii and Dziewonski 2002):

1. The IMIC could be a fossil from an early period of rapid differentiation of the Earth, with the rest of the inner core forming slowly later; in this case, the IMIC could be chemically distinct from the rest of the inner core.
2. The inner core affects the flow pattern in the outer core, and this flow pattern could have changed character when the inner core reached a certain size, altering the anisotropy of the iron core thereafter.
3. The different anisotropy could simply represent a different phase of iron at the pressure and temperature of the IMIC.

After we discuss the heat flow and internal temperatures (Chapter 6) and the nature of the material in the interior of the Earth (7), we will revisit the structure of the Earth and, later, the other terrestrial planets.

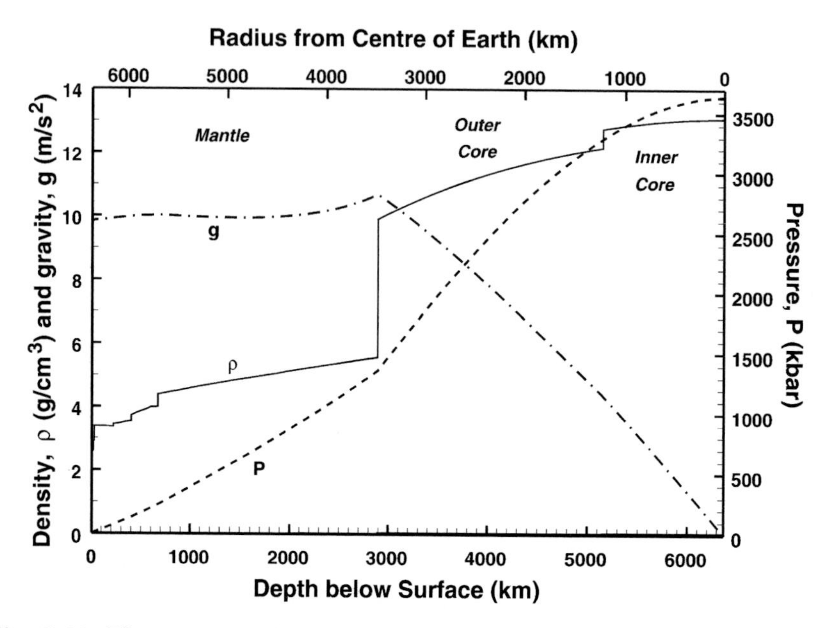

Fig. 5.23. The variation of density, pressure, and gravitational acceleration with radius and depth, according to the preliminary reference Earth model. Produced from data in Dziewonski and Anderson (1981, p. 27)

5.5 Planetary Surfaces

5.5.1 Impacts

The properties of the surfaces of the rocks reveal something of the history of the planet, especially when that surface has been unmodified for eons—like the surfaces of the Moon or Mercury. The extensive cratering on these surfaces shows the effect of a long period of intense bombardment, which the Earth, Venus, and Mars certainly did not avoid, and may have been more intense for them because of their greater gravitational attraction. The small population of colliders that we know as meteors ablates in the atmospheres of the Earth or Venus, but much larger objects—meters or larger in diameter—explosively dissipate much or all of their material. The energy per unit mass arises ultimately from the gravitational potential of the target body but more directly from the relative speed at impact:

$$E = v^2/2 \tag{5.30}$$

The surfaces of Venus, Earth, and Mars have shown extensive modification since a period of intense bombardment in the solar system, some 4 Gy ago.

In the case of Mars, as we shall see, the modification has been selective; on Earth and Venus it has been extensive.

The effect of an impact depends critically on the velocity of the impact and the mass of the impactor. The mass of a meteoroid is unknown typically, but if its size can be determined, an estimate of its density leads to a mass. The density ranges from 1000 kg/m^3 for a "rubble pile" asteroid (an aggregation of loosely packed material) to solid nickel–iron, ~8000 kg/m^3. The speed of impact depends on the orbit of the impactor. If a meteoroid travels in a parallel path to the planet and has in effect no net speed with respect to it, the meteoroid will fall to the planet's surface with the escape velocity of that planet:

$$v_\infty = [2GM_\mathrm{p}/R_\mathrm{p}]^{\frac{1}{2}} \tag{5.31}$$

This quantity is 11.2 km/s for the Earth and only 5.01 km/s for Mars. A meteoroid of asteroidal origin is likely to have originated in the asteroid belt between Mars and Jupiter, although there is a relatively small population of objects even within the orbit of the Earth. There is also a considerable population of objects in the outer solar system, the "trans-Neptunian objects," which are more or less coplanar to the planets and travel in CCW orbits. These include the icy bodies of the Kuiper Belt. Finally, far beyond the 100 or so au of this region is the spherically distributed Oort Cloud, from which we get the long-period comets. A cometary object with a semi-major axis, a, of 10,000 au and perihelion distance, q, of 1 au has an eccentricity (cf. Chapters 2.2 and 3.2):

$$e = 1 - q/a = 1 - 10^{-4} \approx 1 \tag{5.32}$$

Therefore the speed of the comet at perihelion is just short of the escape velocity from 1 au. The escape speed is:

$$v_\mathrm{esc} = [2GM_\odot/q]^{\frac{1}{2}} = 4.21 \times 10^4\,\mathrm{m/s} = 42.1\,\mathrm{km/s} \tag{5.33}$$

If the orbit is retrograde, the velocity is additive to the planet's approximate speed, so that, for an encounter with Earth, the comet would impact with a speed of

$$42.1 + 30.0 + 11.2 = 83.3\,\mathrm{km/s}$$

More generally, the *vis-viva equation* would be used to compute the speed at some point in the orbit and the net speed would be obtained by vectorial addition. Here, however, we have computed a maximum case. The energy per unit mass involved in such a collision would be, from (5.30),

$$E = v^2/2 = 3.5 \times 10^9\,\mathrm{J/kg}$$

for Earth impact. Because the chemical energy released in a TNT explosion is 4.2×10^6 J/kg, such a cometary impact would be equivalent to ~1000 kg of TNT per kg of impactor mass. Of course, small impactors (meters across or less) ablate as they are passing through a planetary atmosphere, and fragments fall to the surface at the terminal velocity, not the escape velocity. Large objects (hundreds of meters to tens of kms, on the Earth), on the other hand, will not be slowed down very much and will impact with great violence, resulting in a very large crater and substantial deposition of material from the conical sheet that is excavated by the impact over the rim and at the center. The material at the forward edge and at ground zero will be vaporized by the high temperatures resulting from the transfer of kinetic energy; the high vapor pressure will then cause a recoil on the trailing edge of the material, which may break up and become distributed over a wide area or even be re-ejected.

In a planetary atmosphere, the material may be carried around the planet, eventually falling to the surface.

We will discuss the roles played by impacts on the surfaces of the planets and moons of the solar system as we discuss them in later chapters (Chapter 8 for the Moon, Chapter 9 for the surfaces of the individual terrestrial planets, and Milone & Wilson (2008, Chapter 13) for the other moons of the solar system).

5.5.2 Observing Planetary Surfaces

5.5.2.1 Phase and Visibility The *phase*, q, of a planet is the fraction of a planet's diameter that appears illuminated by the Sun as viewed from the Earth (or any other platform from which you are viewing the planet!—for now we assume that the Earth is the only viewing site).

The *phase angle*, ϕ, is the angle at the center of the planet between the directions to the Sun and to the Earth. The relation between q and ϕ is:

$$q = {}^1\!/_2\,(1 + \cos\,\phi) \qquad (5.34)$$

See Figure 5.24 for the geometry needed to derive this equation.

Note that when $\phi = 0°$, the planet is fully illuminated (this is a possible configuration for all planets but for practical reasons only an exterior planet can be well viewed when fully illuminated, which for those planets occurs at opposition).

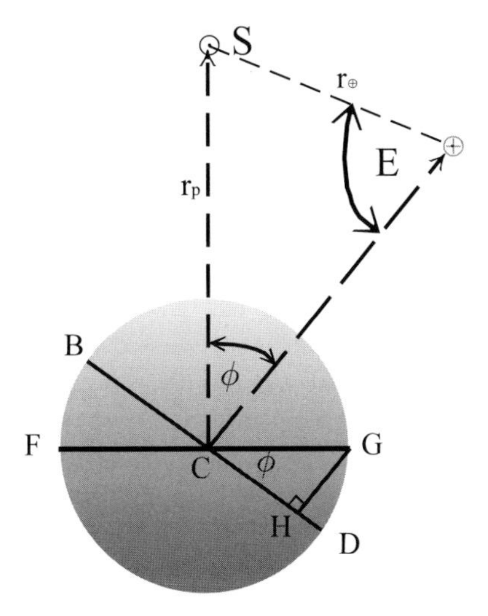

Fig. 5.24. Definition of phase angle

When $\phi = 90°$, half the planet appears illuminated. At a phase angle of 180°, $q = 0$. This is the case only at inferior conjunction, and is possible only for an interior planet.[6]

Although all phase angles are possible for an interior planet (whether easily viewable or not!), this is not true for exterior planets. In those cases, $0° \leq \phi < 90°$, but even sharper constraints can be found, so that $\phi_{\max}$ and $q_{\min}$ may be specified for a given planet.

The sine law for plane triangles allows the phase angle of a planet to be calculated from the *elongation*, E (the angle between the Sun and the planet measured at the Earth, $\sphericalangle S{\oplus}C$ in Figure 2.4), and the distances of the Earth from the Sun, $r_\oplus$, and of the Sun from the planet, r_p, respectively, at any instant.

$$\sin\phi / r_\oplus = \sin E / r_\mathrm{p} \tag{5.35}$$

[6] *Interior*, that is, to the Earth's orbit. Classically, Mercury and Venus are "inferior planets," because they orbit below the orbit of the Sun in the Ptolemaic, geocentric universe. Similarly, planets *exterior* to the Earth's orbit, Mars on out, were classically referred to as "superior planets," because their orbits lay above that of the Sun. In modern usage, "inferior" and "superior" have acquired heliocentric meanings and are used interchangeably with "interior" and "exterior" respectively. See Chapter 1 and Kelley and Milone (2005) for details, configurations, and terminology.

From (5.35), for given values of $r_\oplus$ and r_p, the maximum phase angle, ϕ_{max} (and therefore minimum phase, q_{min}) of a superior planet occur at *quadrature* ($E = 90°$):

$$\sin \phi_{max} = \frac{r_\oplus}{r_p} \tag{5.36}$$

From (5.35) and (5.36), for superior planets, $\phi_{max} \to 90°$ only if $r_p \to r_\oplus$. Mars has the smallest orbit beyond the Earth's and therefore attains the maximum departure from full phase that we see among the superior planets. The maximum possible value of ϕ_{max} for Mars occurs on the extremely rare occasion when the Earth is at aphelion at the same instant that Mars at *quadrature* ($E = 90°$) is at perihelion. From (5.36),

$$\phi_{max} = \arcsin\,(1.017/1.381) = 47.4°$$

which, when inserted in (5.34), leads to a minimum possible phase, $q_{min} = 0.838$. Because of the low probability of the required conditions occurring, observed phases of Mars will almost always be greater than this.

5.6 Addendum: Properties of Legendre Polynomials and Associated Legendre Functions

In simpler form, the function of interest in (5.7), namely,

$$\sum_{n=0}^{\infty} P_n(\cos\,\theta)\left(\frac{a}{r}\right)^n = \left[1 - 2(\cos\,\theta)\left(\frac{a}{r}\right) + \left(\frac{a}{r}\right)^2\right]^{-1/2}$$

$$= \sum_{n=0}^{\infty} \frac{(2n)!}{2^{2n}(n!)^2}\left[2(\cos\,\theta)\left(\frac{a}{r}\right) - \left(\frac{a}{r}\right)^2\right]^n \tag{5.37}$$

is convergent for $(a/r) < 1$. The *Legendre polynomial* part of this function, $P_n(\cos\,\theta)$, is written as:

$$P_n(x) = \sum_{k=0}^{[n/2]} (-1)^k \frac{(2n-2k)!}{2^n k!(n-k)!(n-2k)!} x^{n-2k} \tag{5.38}$$

where $x \equiv \cos\theta$, and the upper limit on the summation $[n/2] = n/2$ for even, and $(n-1)/2$ for odd, n. Whence, $P_0(x) = 1$, $P_1(x) = x$, $P_2(x) = \frac{3}{2}x^2 - \frac{1}{2} = \frac{1}{2}(3x^2 - 1)$, $P_3(x) = \frac{1}{2}(5x^3 - 3x)$, and, generally,
$(2n+1)xP_n(x) = (n+1)P_{n+1}(x) + nP_{n-1}(x)$, $n = 1, 2, 3\ldots$. Also, $\partial P_{n+1}/\partial x = (\mathrm{n}{+}1)P_n(x) + x\,\partial P_n(x)/\partial x$.

Further, we can write,

$$P_n(x) = \frac{1}{2^n n!} = \left(\frac{\mathrm{d}^n}{\mathrm{d}x^n}\right)(x^2 - 1)^n \tag{5.39}$$

Finally, where "$m = 0/1$" designates "either 0 or 1," $P_n(x)$ can be written

$$P_n(\cos\,\theta) = \sum_{m=0/1}^{n} a_m \cos\,m\,\theta = \sum_{m=0/1}^{n} a_m(\mathrm{e}^{\mathrm{i}m\theta} + \mathrm{e}^{-\mathrm{i}m\theta}) \tag{5.40}$$

The *Associated Legendre functions* are characterized by:

$$P_n{}^m(x) = (1 - x^2)^{m/2}\left(\frac{\mathrm{d}^m}{\mathrm{d}x^m}\right)P_n(x) \tag{5.41}$$

where $m \leq n$. Thus, $\mathrm{P}_1{}^1(x) = (1 - x^2)^{1/2} = \sin\,\theta$; $P_2{}^1(x) = 3x(1 - x^2)^{1/2} = 3\,\cos\,\theta \cdot \sin\,\theta$, $P_2{}^2(x) = 3\,\sin^2\,\theta, \ldots$.

References

Beghein, C. and Trampert, J. 2003. "Robust Normal Mode Constraints on Inner-Core Anisotropy from Model Space Search," *Science*, **299**, 552–555.

Boehler, R. 2000. "High-Pressure Experiments and the Phase Diagram of Lower Mantle and Core Materials," *Reviews of Geophysics*, **38**, 221–245.

Calvet, M., Chevrot, S., and Souriau, A. 2006. "P-Wave Propagation in Transversely Isotropic Media II. Application to Inner Core Anisotropy: Effects of Data Averaging, Parametrization and A Priori Information," *Physics of Earth and Planetary Interiors*, **156**, 21–40.

Consolmagno, G. J. and Schaefer, M. W. 1994. *Worlds Apart* (London: Prentice Hall).

Cox, A. (ed.), 2000. *Allen's Astrophysical Quantities*, 4th Ed. (New York: Springer-Verlag)

Dziewonski, A. M. and Anderson, D. L. 1981. "Preliminary Reference Earth Model," *Physics of the Earth Planetary Interiors*, **25**, 297–356.

Ishii, M. and Dziewonski, A. 2002. "The Innermost Inner Core of the Earth: Evidence for a Change in Anosotropic Behavior at the Radius of about 300 km," *Proceedings, National Academy of Science*, **99**, 14026–14030.

Ishii, M., and Dziewonski, A. 2003. "Distinct Seismic Anisotropy at the Centre of the Earth," *Physics of Earth and Planetary Interiors*, **140**, 203–217.

Kelley, D. H. and Milone, E. F. 2005. *Exploring Ancient Skies: An Encyclopedic Survey of Archaeoastronomy* (New York: Springer).

Milone, E. F., and Wilson, W. J. F. 2008. *Solar System Astrophysics: Planetary Atmospheres and the Outer Solar System.* (New York: Springer).

Smith, D. G. (ed.) 1981. *The Cambridge Encyclopedia of Earth Sciences* (New York: Crown Publishers/Cambridge University Press).
Stacey, F. D. 1969. *Physics of the Earth* (dia of Earth Sciences (New York: Crown Publishers/Cambridge University Press).
Stacey, F. D. 1969. *Physics of the Earth* (New York: Wiley).
Williams, Q., Jeanloz, R., Bass, J., Svendsen, B., and Ahrens, T.J. 1987. "The Melting Curve of Iron to 250 Gpa: A Constraint on the Temperature at the Earth's Center," *Science*, **236**, 181–182.
Zharkov, V. N. and Trubitsyn, V. P. 1978. *Physics of Planetary Interiors* (Tucson, AZ: Pachart Publishing House). Trs. & ed. E. B. Hubbard.

Challenges

[5.1.] Derive equation (5.3b).

[5.2.] Assume that a meteroid originates from a direct orbit with $a = 1.333$ au and $e = 0.25$, and collides with the Earth. Compute its speed at 1 au and on impact with Earth. If the object has a diameter of 100 m and a mean density of $3300\,\mathrm{kg/m^3}$, compute (a) the energy per unit mass and (b) the total energy of the explosion. (c) What do you suppose will happen if the object strikes an ocean rather than land? Will it leave a crater, for instance?

[5.3.] Derive equation (5.34) and demonstrate why an exterior planet cannot be observed at phase angles $\leq 90°$. Show also that the maximum observed phase angle for an exterior planet occurs when the planet is at quadrature.

[5.4.] Prove that the phase is a minimum when the phase angle is a maximum. Compute the minimum observable phases for the asteroid Ceres, Jupiter, and Pluto. (Orbital data are available in Milone & Wilson (2008, Chapters 15, 12, and 13, respectively.))

6. Planetary Heat Flow and Temperatures

The surface temperatures of planets in our solar system currently depend basically on four quantities:

1. The luminosity of the Sun
2. The distance of the planet from the Sun
3. The planetary bolometric albedo
4. The heat welling up from the interior

The first two of these determine the solar energy flux reaching the planet (the planet's solar constant) and the third determines the energy flux actually absorbed by the planet. In the early solar system, energy from frequent large impacts was a major contribution to planetary heat budgets. At present, it is not (although some of the heat energy accumulated from that source is still retained in the deep interiors of all the planets), so only the above four factors will be discussed here. A planet's surface temperature is then determined by the equilibrium condition that the energy absorbed from the Sun and the energy welling up from the interior must together equal the energy radiated by the planet. The total emitted radiation of planets generally exceeds that absorbed from the Sun, some by significant amounts. Saturn, for example, radiates 78% more heat than it receives, the excess coming from internal heat sources. On the other hand, the internal heat sources in the terrestrial planets are far less important than solar radiation, as we demonstrate below for the Earth.

To make a long story short, the average heat flux from Earth's interior is $0.082\,\mathrm{W/m^2}$, and the total power radiated by the interior sources is 4.2×10^{13} W. This compares to a radiative power of 1.09×10^{17} W absorbed from the Sun.

First we consider the heat flow from the interior, and the resulting geotherms for the Earth. In later chapters we will deal with the other terrestrial planets in a comparative way and, separately, with the gas giants.

6.1 Heat Flow

The average observed heat flux from the interior is (Figure 6.1):

$$F_{\mathrm{AVG}} = 82\,\mathrm{mW/m^2} = 0.082\,\mathrm{W/m^2}$$

The total heat energy escaping from Earth per second is then:

$$\begin{aligned} L &= 4\pi R^2 F_{\mathrm{AVG}} \\ &= (4\pi)(6378 \times 10^3\,\mathrm{m})^2 \\ &\quad \times (0.082\,\mathrm{W/m^2}) \\ &= 4.2 \times 10^{13}\,\mathrm{W} \end{aligned}$$

6.1.1 Sources of Internal Heat

The sources of heat can be classified into two types:

1. *Primordial heat.* Heat produced during the formation of the Earth; the Earth is still cooling down as a result.
2. *Ongoing processes.* Heat being generated inside the Earth now.

We summarize the evidence for each in point form.

6.1.1.1 Primordial Heat

Primordial heat has several likely sources, separately or in combination.

(1) Accretional Heat

Figure 6.2 illustrates impacts during early accretion. The source is the heat generated by the sudden stopping of incoming planetesimals during the accretion and later bombardment of the early Earth. Such events caused partial melting of the earth.

(2) Possible Moon-Forming Impact

Figure 6.3 illustrates this major event (see Chapter 8.5). This event had three main features:

- ~Mars-sized impactor struck the Earth off-center.
- Core sank into the Earth's core.

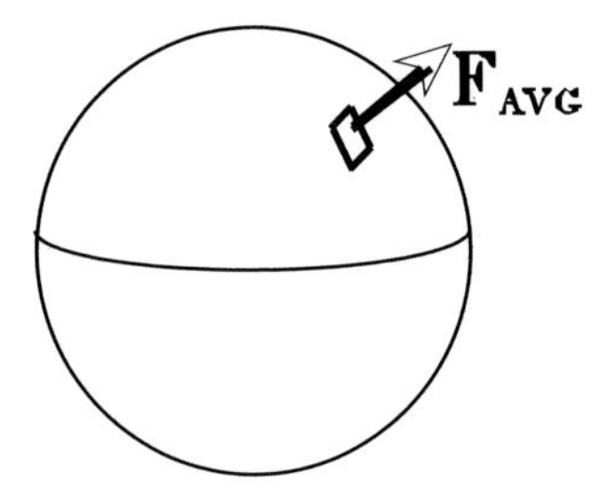

Fig. 6.1. Earth's heat flux

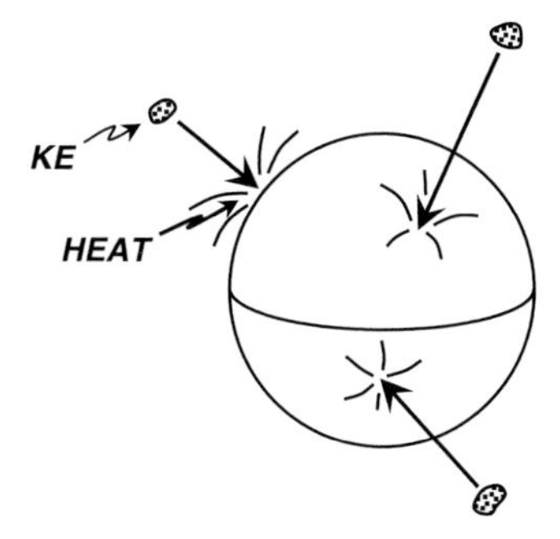

Fig. 6.2. Accretion heating

- Mantle and part of the Earth's mantle were expelled to form the Moon.

(3) Short-Lived Radioactive Isotopes

- Radiogenic heating caused, for example, by $^{26}\text{Al} \longrightarrow {}^{26}\text{Mg}$ (half-life $t_{1/2} = 7.4 \times 10^5$ y).
- The energy released through the decay heats the Earth.
- With a half-life of 7.4×10^5 y, ^{26}Al is no longer left in the Earth. Most of the decay and heating took place during the first few half-lives (say 5 million years), so this is a component of the primordial heat.

(4) Differentiation of the Earth

Figure 6.4 illustrates the potential energy (PE) of droplets of liquid iron as they trickle down to create the core.

- PE is converted to kinetic energy (KE) in the fall toward the core.
- KE is converted into heat by friction and "collisions" with surrounding rock.

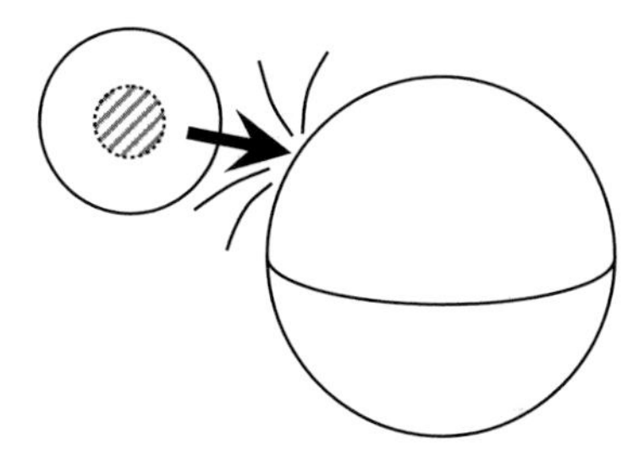

Fig. 6.3. Impact heating in a major event

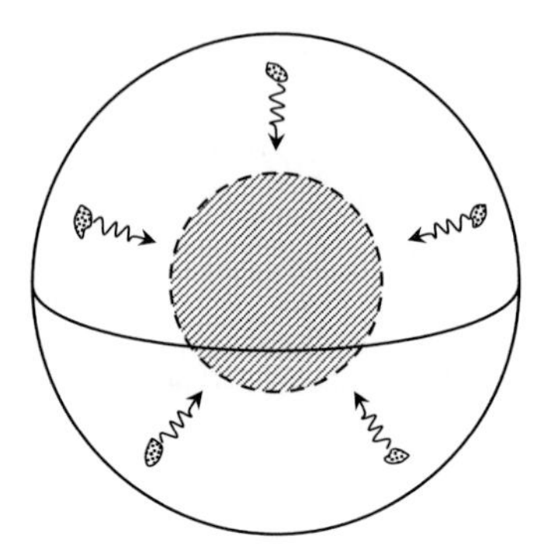

Fig. 6.4. Trickle-down heating from differentiation

- The net result is that gravitational energy is released to form heat.
- The core was formed during the first 0.5 billion years, therefore this source is included as a component of the primordial heat of the Earth.

(5) Present-Day Heat Flow due to Primordial Heat

It is difficult to estimate what fraction of the total present-day heat flow from the Earth is due to any or all of the primordial heat sources discussed above. One carefully developed heat flow model suggests 15–35%, whereas another gives 7–12%. In either case, however, the contribution is significant and should not be ignored.

6.1.1.2 Ongoing Processes

Ongoing processes include several probable sources.

(1) Radiogenic Heat with Very Long Half-Lives

The main sources are:

$$
\begin{aligned}
{}^{235}\mathrm{U} &\rightarrow \cdots \rightarrow {}^{207}\mathrm{Pb} \\
{}^{238}\mathrm{U} &\rightarrow \cdots \rightarrow {}^{206}\mathrm{Pb} \\
{}^{232}\mathrm{Th} &\rightarrow \cdots \rightarrow {}^{208}\mathrm{Pb} \\
{}^{40}\mathrm{K} &\rightarrow {}^{40}\mathrm{A} \text{ or } {}^{40}\mathrm{Ca}
\end{aligned}
$$

- Radiogenic heat is the most important individual source of heat in the Earth.
- It is hard to measure the total amount of U, Th, K in the Earth, but most estimates give a heat flux in the range $(2.1 \pm 0.6) \times 10^{13}$ W or $50\% \pm 15\%$ of the total heat flux from the Earth.

(2) Latent Heat Released in the Core

- This is the heat released by solidification and crystallization of the molten iron in the outer core, as it cools.

(3) Gravitational Energy Release in the Core

This has at least two components in the Earth:

- The iron in the core shrinks as it cools.
- Iron crystals fall inward to form the solid, inner core. (The mantle also falls downward onto the shrinking iron core.)

The total heat flow from the core to the mantle, including possible radioactive decay (^{40}K) is ~20% of total heat flux from the Earth.

6.1.2 Methods of Energy Transport

Energy can be transported in any of three ways: radiation, conduction, and convection.

6.1.2.1 Radiation

Rock is opaque, so radiation is not significant.

6.1.2.2 Conduction and Convection

Conduction involves the slow diffusion of heat from a warmer region to a cooler region, whereas convection involves the physical transport of the warmer material to a new site. Physical transport is faster than diffusion, so when convection occurs it is very efficient and by far dominates conduction.

A test for convection is provided by the *Rayleigh number*:

$$Ra = \frac{\text{buoyancy force (favoring convection)}}{\text{viscous drag force (hindering convection)}}$$

The critical Rayleigh number for convection is $Ra_{\mathrm{c}} \sim 2000$, where the subscript c stands for "critical." The condition is then,

$$(Ra > Ra_{\mathrm{c}} \Rightarrow \text{convection})$$

The Rayleigh number for a fluid heated from within is given by:

$$Ra = \frac{g\,\alpha\,q\,h^5}{\nu\,\kappa\,k} \qquad (6.1)$$

From this we can estimate the value of Ra for the mantle as a whole, given the following properties:

ρ = density $\sim 4\,\text{g/cm}^3 = 4 \times 10^3\,\text{kg/m}^3$
α = thermal expansion coefficient $\sim 3 \times 10^{-5}/\text{K}$
κ = thermal diffusivity $\sim 10^{-6}\,\text{m}^2/\text{s}$
g = acceleration due to gravity $\sim 10\,\text{m/s}$
ν = coefficient of kinematic viscosity $\sim 3 \times 10^{17}\,\text{m}^2/\text{s}$
h = thickness of layer $\sim 3000\,\text{km} = 3 \times 10^6\,\text{m}$
k = thermal conductivity $\sim 4\,\text{Wm}^{-1}\,\text{K}^{-1}$
q = rate of internal heat production $\sim 9 \times 10^{-12}\text{W/kg}$

Then,

$$Ra \sim 5 \times 10^5 >> Ra_c \Rightarrow \text{Convective.}$$

Therefore, heat is transported out through the mantle by convection because warmer rock is buoyant and rises; cooler rock is denser and sinks. Under pressure, the rock in the mantle behaves like a high-viscosity fluid and undergoes plastic deformation (like a glacier). The speeds of convective motion are ~cm per year.

Convection can be either laminar (smooth flow) or turbulent. A useful parameter to determine which applies is the *Reynold's number*:

$$Re = \frac{ud}{\nu} \tag{6.2}$$

where u is the speed of flow, d is the thickness of the convecting layer, and ν is the kinematic viscosity. The flow is laminar if $Re << 1$ and turbulent if $Re >> 1$. The best estimates for the Reynold's number in the mantle are around 10^{-21} (Anderson, 1989, p. 255), in agreement with that calculated with the data above, so the flow is almost certainly laminar.

In the crust, heat transport is by conduction.

6.1.2.3 Nature of the Convection

There are two major possibilities:

(1) Whole-Mantle Convection

Convection mixes the material; therefore, this model implies that the upper and lower mantles have basically the same composition, although segregation of rising rock into basaltic magma and olivine/pyroxene solids still takes place in the upper mantle.

With this model, the 400 and 650 km discontinuities must be due to phase transitions only, and not composition discontinuities.

Figure 6.5 illustrates this case. It also shows two features of the continental drift theory of modern geology: the uplift and spreading at a crustal plate interface and the subduction of a plate at another. Note that the descending portion of the crust, subducting under the thicker continental plate at the left, is free, in this model, to be reabsorbed fully into the deepest part of the upper mantle.

(2) Two-Layer Convection

In this case, illustrated in Figure 6.6, the lower mantle and the upper mantle convect separately. The lower mantle convection drives the upper mantle convection, but the two are not *mechanically* coupled. There is a thermal boundary layer separating them at the 650 km discontinuity. This layer is non-convecting, with heat transport across it being by conduction.

Hot rock rising in the lower mantle creates a hot spot in the boundary layer. This in turn heats the rock at the base of the upper mantle, which then rises and drives the upper mantle convection. In this case, the 650 km discontinuity could also be a composition discontinuity, as well as a phase transition.

The lower mantle may be more silica-rich (more $MgSiO_3$ perovskite, with Mg: Si $\sim$ 1:1) than the upper mantle (more Mg_2SiO_4 olivine, with Mg: Si $\sim$ 2:1). The lower mantle may also be more iron-rich than the upper mantle (possibly entering the lower mantle through the *D'' layer* that separates the lower mantle from the core).

Each model for mantle convection has strong supporters, and the question is not settled.

We consider next the variation of temperature as a function of depth within the Earth: $T = T(z)$.

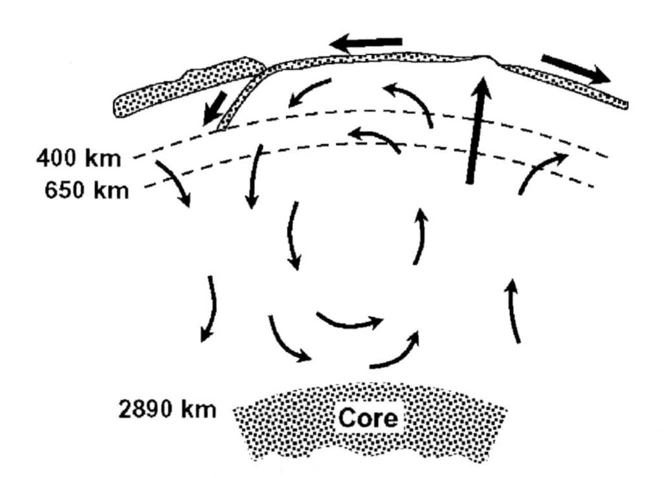

Fig. 6.5. Whole-mantle convection

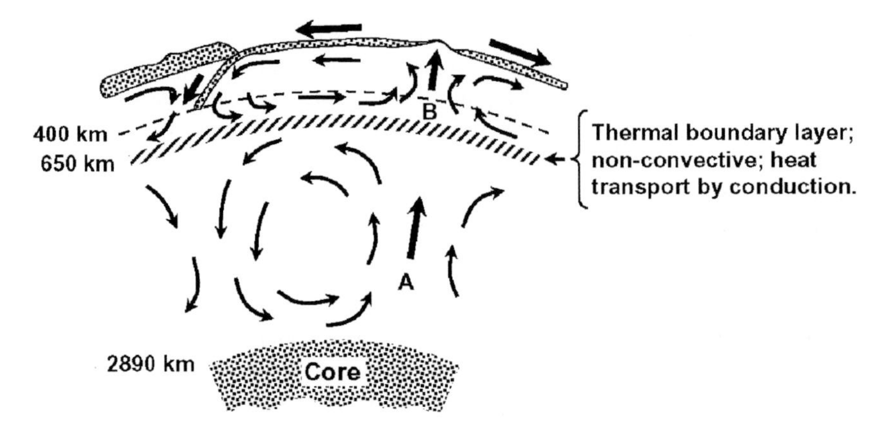

Fig. 6.6. Two-layer mantle convection

6.1.2.4 Temperature Gradient Figure 6.7 shows a horizontal slab of rock whose top surface is at a depth z in the Earth and whose bottom surface is at a depth $z + \Delta z$.

Here we adopt the convention that the depth, z, is zero at the Earth's surface and increases downward. That is, both z and Δz are positive downward.

The temperature of the top surface is T and that of the bottom surface is $T + \Delta T$. We expect temperature to increase inward in the Earth, so ΔT will be positive when Δz is positive. The ratio

$$\frac{\Delta T}{\Delta z}$$

gives the rate of change of temperature with depth at any fixed instant of time, and is called the *temperature gradient*. Because ΔT and Δz are both positive downward, it follows that the temperature gradient is positive downward.

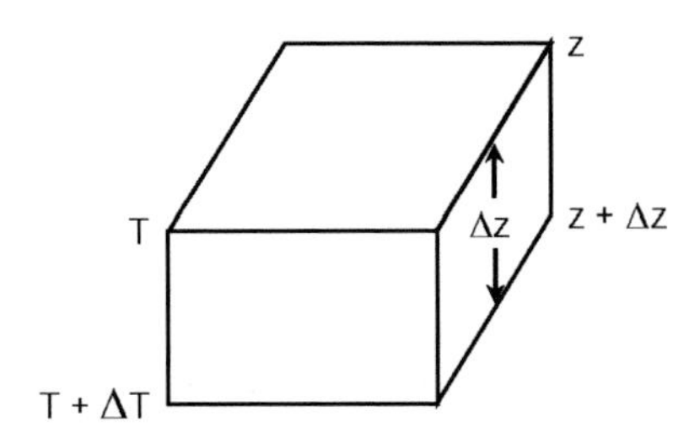

Fig. 6.7. Defining a temperature gradient

In the limit as $\Delta z \to 0$, we have

$$\lim_{\Delta z - 0} \frac{\Delta T}{\Delta Z} = \frac{\partial T}{\partial Z} \tag{6.3}$$

The use of partial derivative notation on the right side of (6.3) means that we are taking the derivative of T with respect to z while holding time fixed; that is, we are comparing the temperatures at two different points at the same instant of time.

By way of a counterexample, suppose a piece of rock is moving downward, as in mantle convection. This situation is illustrated in Figure 6.8.

Then the total derivative,

$$\frac{\mathrm{d}T}{\mathrm{d}z}$$

gives the rate of change of T with z for this rock as it moves; that is, the temperature is being compared at two different points at two *different* instants of time (the instants when the rock occupies those two points). So the full derivative does not depend on the depth alone, but also on time; while the partial derivative with respect to the depth depends only on the depth. In the language of mathematics, Figure 6.8 demonstrates that $\mathrm{d}T/\mathrm{d}z$ depends on z and t.

6.1.3 Heat Conduction

Energy transport in the Earth's crust is by conduction.

Define the heat flux, Q, as the amount of heat energy passing through each square meter of surface area per second. The units of heat flux are therefore $\mathrm{J\,m^{-2}\,s^{-1}}$ or $\mathrm{W/m^2}$.

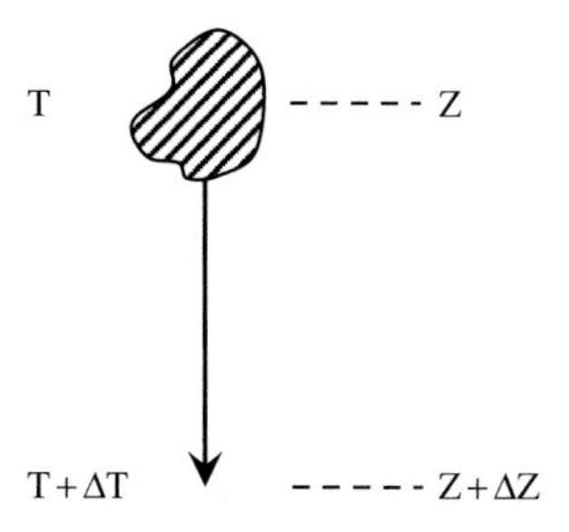

Fig. 6.8. Convective temperature gradient

Q is defined to be positive in the direction of increasing z, viz., downward, in the convention we have adopted here (see Figure 6.9).

It turns out that, for a fixed separation Δz, $|Q|$ is larger for a larger temperature difference, ΔT; but, for a given ΔT, $|Q|$ is smaller if the surfaces are further apart (Δz is larger):

$$|Q| \propto \Delta T, \text{ but inversely} \propto \Delta z$$

Therefore, $|Q| \propto \dfrac{\Delta T}{\Delta z}$; and, in the limit as $\Delta z \to 0$,

$$|Q| \propto \frac{\partial T}{\partial z}$$

The constant of proportionality, k, is defined such that:

$$Q = -k\frac{\partial T}{\partial z} \tag{6.4}$$

where k is the *coefficient of thermal conduction*, and the minus sign is there because heat always flows from the hotter region toward the cooler region; that is, in the direction opposite to the temperature gradient (as shown in Figure 6.9). The units of k are given by $Q/(\Delta T/\Delta z)$, i.e., $(\mathrm{W/m^2})/(\mathrm{K/m}) = \mathrm{W\,m^{-1}\,K^{-1}}$.

Equation (6.4) is the one-dimensional form of a more general expression known as *Fourier's Law.* Equation (6.4) will, however, suffice for our treatment here. This is our *thermal conductivity equation.* A typical value of Q for crustal rocks on the Earth's surface is $\sim 3\,\mathrm{W\,m^{-2}\,K^{-1}}$.

6.1.4 Energy Generation

Heat energy is generated inside the volume of rock because the radioactive decay of uranium, thorium, etc., releases heat. There may also be other

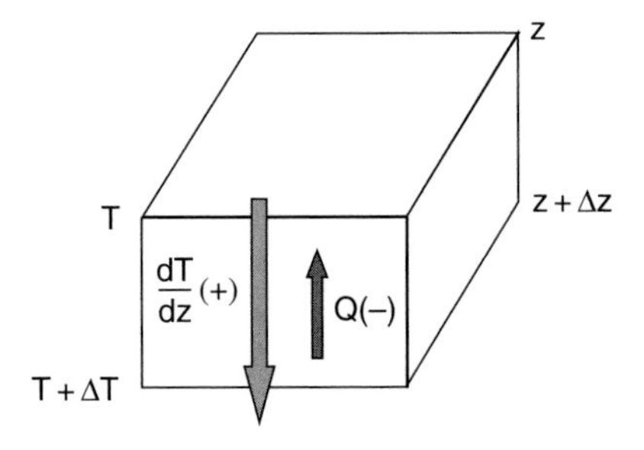

Fig. 6.9. Sign convention for Q and $\mathrm{d}T/\mathrm{d}z$

Color Plates

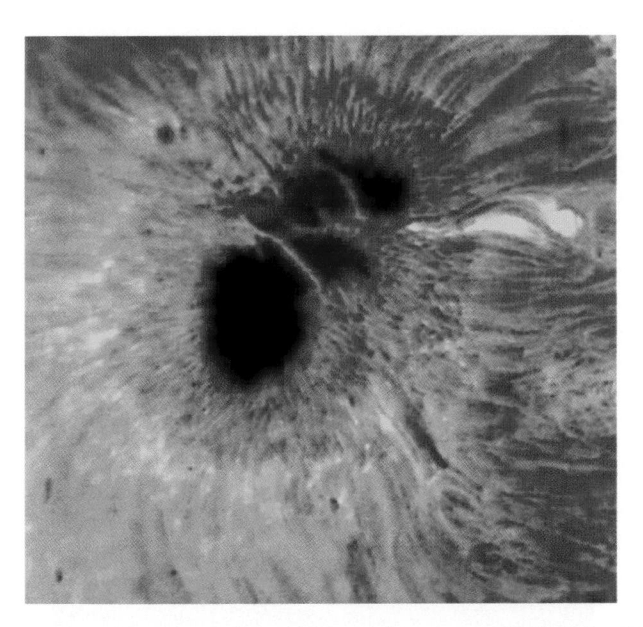

Plate 1. An active sunspot group (AR 8971) as recorded in the Hα spectral line on April 27, 2000, with the McMath–Pierce Solar Telescope at the Kitt Peak National Observatory, Arizona. The area covered is 100 $\times$ 100 arc-secs. Courtesy Dr T. A. Clark. The McMath–Pierce Solar Telescope is operated for the National Science Foundation by the Association of Universities for Research in Astronomy as part of the National Solar Observatory.

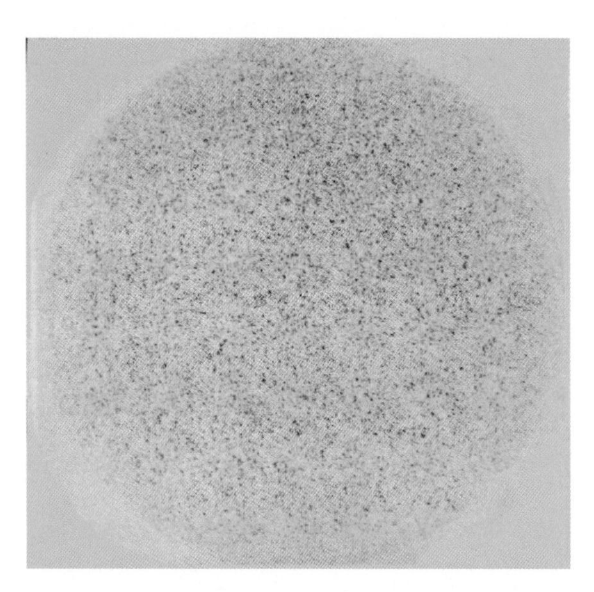

Plate 2. Illustration reprinted with permission from Harvey (1995, Fig. 1). The Doppler signature of hot rising (*red*) and falling (*blue*) gas from two images obtained 2 min apart at an observatory at the South Pole in January, 1991. Copyright 1995, American Institute of Physics.

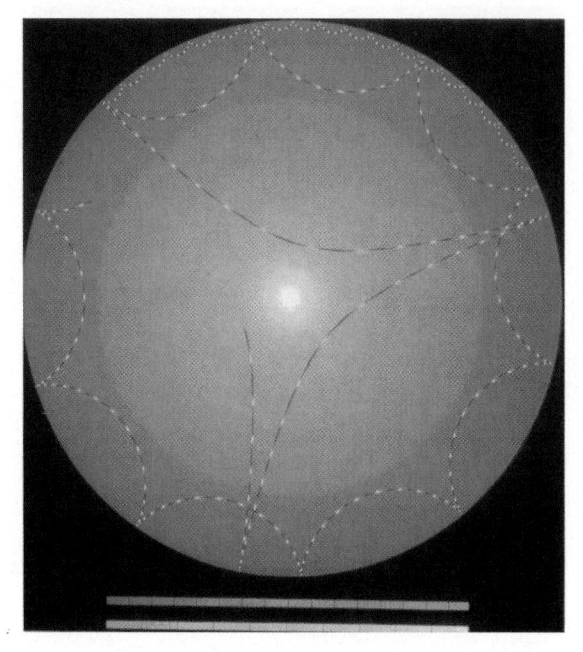

Plate 3. A cutaway model of the Sun, showing refracted acoustic rays (*dashed curved lines*) and the temperature (*color, increasing from left to right along the legend bar*) and energy production (intensity) through the solar interior. Ray calculations were done by Dr. D. D'Silva. Illustration reprinted with permission from Harvey (1995, Fig. 2). Copyright 1995, American Institute of Physics.

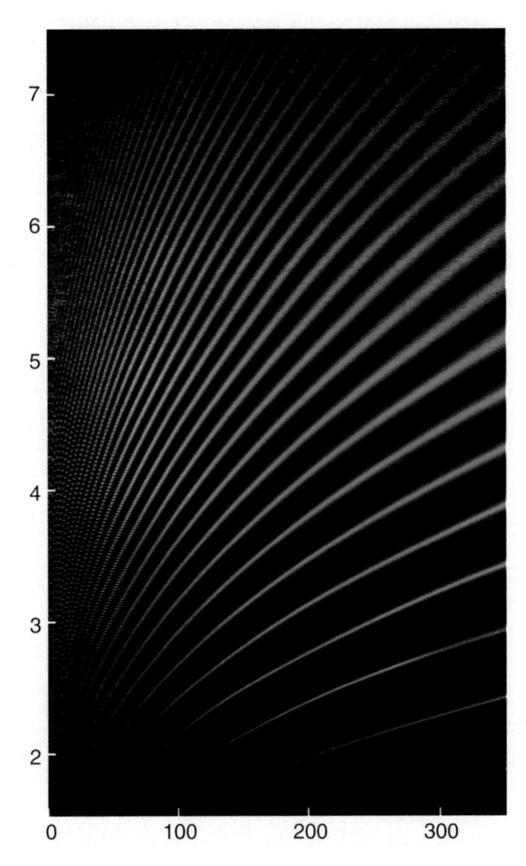

Plate 4. The power spectrum of solar oscillations in the form of standing acoustic wave patterns. The frequency in mHz is plotted against the spherical harmonic degree, ℓ. In this plot, color represents depths, with *blue* indicating shallow and *red*, indicating deep, penetrating modes. Image supplied by John W. Harvey. Illustration reprinted with permission from Harvey (1995, Fig. 3). Copyright 1995, American Institute of Physics.

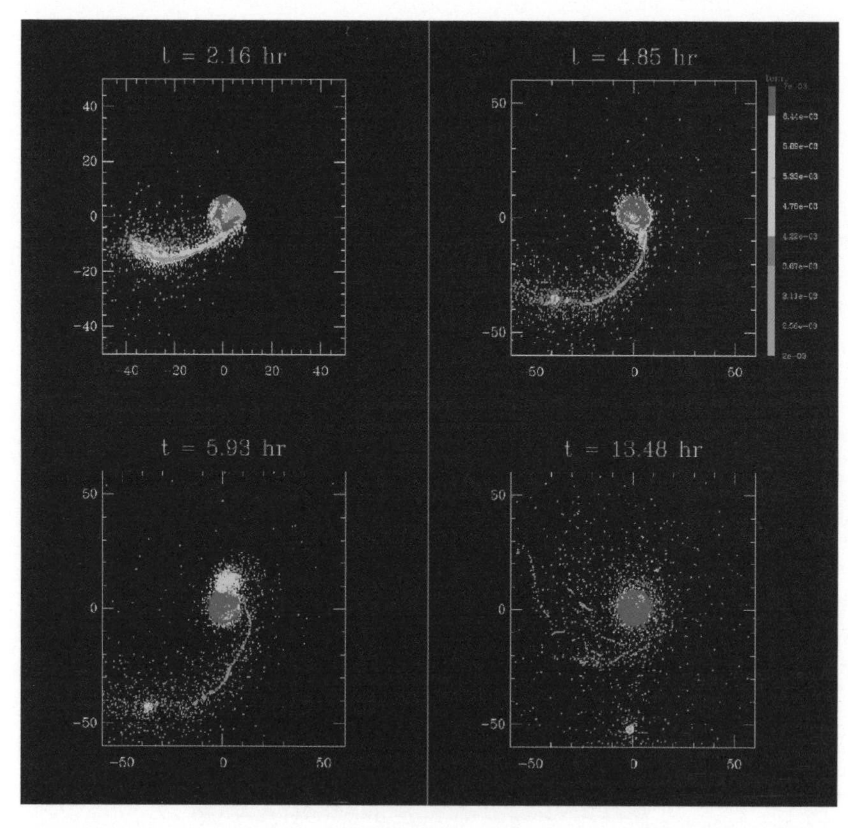

Plate 5. Stages in the coalescence of the Moon after a major impact involving a proto-Earth in the early solar system. Much of the material falls back to Earth, a small amount escapes the system, and some goes into orbit around the Earth to form a disk that later coalesces into the Moon. The clump at 6 o'clock in the lower right panel is on an eccentric orbit that later passes within the Earth's Roche limit (see Milone & Wilson 2008 Chapter 13.2), so that this clump, itself, is not a candidate for the Moon. Note that simulated temperatures vary from ~2000 K for many fragments to ~7000 K at the Earth. From a simulation by Robin Canup, and kindly provided by her (see Canup, 2004 for a full discussion) and appearing here with the permission of Elsevier.

Plate 6. The Mars Exploration Rover *Opportunity* found evidence of a water environment in Endurance Crater in the form of flow channels and "blueberries," shown here. Credits: NASA/JPL/Cornell. Image PIA06692.

Plate 7. A sub-frame of an image from the Mars Reconnaisance Orbiter High Resolution Imaging Science Experiment (HiRISE), in false color to delineate reflectance from different materials. This view of a scarp of Chasma Boreale, near the north pole of Mars shows layered deposits overlying darker material. Note the interweaving of bright, ice-laden layers with darker sand layers. NASA/JPL-CalTech/University of Arizona image PIA09097.

sources or sinks of heat, such as pressure changes, latent heat released by phase changes, etc.

Define A = heat energy generated per cubic meter per second (W/m^3).

6.1.5 Equilibrium

The Earth and the other terrestrial planets have had time to lose most of their heat of formation (primordial heat), so we expect rocks in the crust to be in thermal equilibrium, on average. *Thermal equilibrium* means that rate of energy loss = rate of internal energy generation.

This is actually a quasi-equilibrium condition because, for example, the radioactive decay rate is decreasing very slowly with time; but these changes are slow enough that they can be ignored compared to the timescale of heat conduction.

Each individual volume inside the crust should also be in thermal equilibrium (on average), so the net energy gained by the volume over any length of time, δt, should be zero:

The heat energy through bottom surface (positive if into the volume)
plus
The heat energy through top surface (negative if out of the volume)
plus
the heat generated in the volume
= 0.

Figure 6.10 illustrates the situation.

The individual terms for the heat flux are, for the heat energy up through the bottom:

$$\text{heat energy through bottom} = \begin{bmatrix}\text{heat energy per unit area} \\ \text{per unit time} \times \text{area} \times \text{time}\end{bmatrix}$$
$$= -Q_{\mathrm{B}} a \, \delta t. \tag{6.5}$$

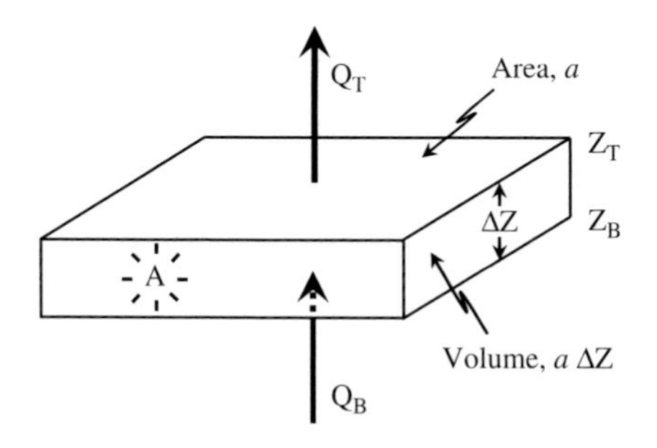

Fig. 6.10. The energy budget in a slab

The minus sign comes from the circumstance that the volume gains energy (LHS positive) with an upward heat flow (Q_{B} negative). The net result is that the quantity $(-Q_{\mathrm{B}})$ is positive.

And, for the heat energy through the top,

$$\text{heat energy through top} = \begin{bmatrix}\text{heat energy per unit area} \\ \text{per unit time} \times \text{area} \times \text{time}\end{bmatrix}$$
$$= Q_{\mathrm{T}} a\,\delta t. \tag{6.6}$$

The volume loses energy (LHS negative) with an upward heat flow (Q_{T} negative), so the sign in the equation is positive.

Now for the heat generated within the volume:

$$\text{Heat generated in } \Delta V = \text{heat generated per unit volume per unit time} \times \text{volume} \times \text{time}$$
$$= A\,\Delta V\,\delta t = A\,a\,\Delta z\,\delta t. \tag{6.7}$$

These three terms have to add up to zero net energy gain, so,

$$-Q_{\mathrm{B}}\,a\,\delta t + Q_{\mathrm{T}}\,a\,\delta t + A\,a\,\Delta z\,\delta t = 0. \tag{6.8}$$

Dividing through by a and δt and rearranging terms gives

$$Q_{\mathrm{B}} - Q_{\mathrm{T}} = A\,\Delta z \tag{6.9}$$

or

$$\frac{Q_{\mathrm{B}} - Q_{\mathrm{T}}}{\Delta z} = \frac{Q_{\mathrm{B}} - Q_{\mathrm{T}}}{z_{\mathrm{B}} - z_{\mathrm{T}}} = A \tag{6.10}$$

Therefore, in the limit as $\Delta z \to 0$,

$$\frac{\partial Q}{\partial z} = A \tag{6.11}$$

If we use the thermal conduction equation (6.4) in (6.11) and divide through by $-k$, then we arrive at

$$\frac{\partial^2 T}{\partial z^2} = -\frac{A}{k} \tag{6.12}$$

Equation (6.12) can be integrated to find T as a function of z in the crust, where heat transport is by conduction, provided A and k are known at depth z.

However, A and k are not obtainable from seismic work in the manner that we can find v_{P} and v_{S}, the P- and S-wave velocities, discussed in Chapter 5. Thus, the geotherms are much more uncertain than the run of density with depth.

6.1.6 Central Temperature of the Earth

If energy were transported by conduction throughout the Earth, integration of (6.12) would be valid and would yield a temperature for the center of the Earth of about 60,000 K. This extremely high temperature results from the fact that conduction is a slow process, so a large temperature gradient is needed to produce the observed heat flow. The faster the temperature increases with depth into the Earth, the higher the central temperature will be.

Convection is a much more efficient process, because the warmer rocks are transported to cooler regions (carrying their heat with them) in a much shorter time than it would take the same heat to diffuse to the new region by conduction. The observed heat flow can then be produced by a much smaller temperature gradient, giving a central temperature for the Earth near 6000–8000 K.

6.2 Geotherms

A *geotherm* is an equation (or a table, or a curve on a graph) which gives temperature as a function of depth into the Earth. This can be created readily if we make certain simplifying assumptions. One of these involves adiabatic conditions. An *adiabatic process* is any process in which heat energy does not enter or leave the material involved. The temperature may change due to compression or expansion, but under adiabatic conditions, there is no exchange of heat energy *between* the material and its surroundings.

The speed of convection in the Earth is high compared to the rate of diffusion of heat by thermal conduction, so heat does not have time to enter or leave the moving material and the convection is adiabatic.

The temperature gradient in material which is convecting adiabatically is

$$\frac{\partial T}{\partial r} = -\frac{T\alpha g}{c_P} \tag{6.13}$$

where α is the coefficient of thermal expansion; c_P the specific heat at constant pressure; g the acceleration due to gravity; and T the temperature.

Integration of (6.12) for conduction gives the temperature and rate of heat flow at the base of the crust (top of the mantle). Equation (6.13) can then be integrated numerically from the top of the mantle to the center of the Earth (the outer core is also convecting), giving a geotherm for the whole Earth.

Typical results are shown in Figure 6.11 where the hatched region shows the uncertainty in the results.

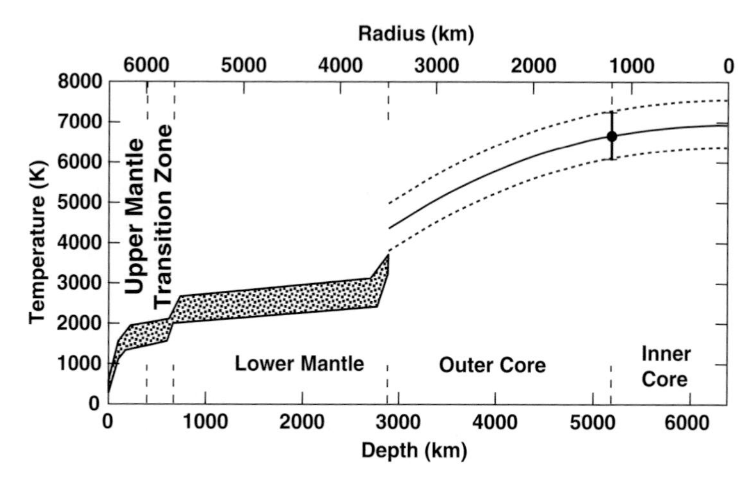

Fig. 6.11. Geotherm for the Earth. Mantle temperatures are from Jeanloz & Morris (1986) and references therein. The geotherm in the core is an approximate interpolation between three points: 4400 K ± 600 K at the core outer boundary (Jeanloz & Morris, 1986); 6670 K ± 600 K at the inner core-outer core boundary (Alfé et al., 1999); and ≤ 6900 K at the Earth's center (Williams et al., 1987), taken here as 6900 K. The first two of these temperatures are consistent with those of Williams et al. (1987): 4800 K ± 200 K and 6600 K, respectively. The temperature gradient is very steep at the core-mantle boundary, and is left as unknown in the figure above. Approximate uncertainties in the mantle temperatures are indicated by the width of the geotherm, and in the core by the dotted lines, based on the uncertainties in the temperatures at the inner and outer boundaries of the outer core. We thank Bukowinski (1999) for the references to this material

The uncertainty results from the uncertainties in α and c_P at each depth. The data are based on laboratory experiments which measure the properties of iron and various rock mixtures at high temperatures and pressures. From this process, the temperature at the center of the Earth is determined to be $6900\,\mathrm{K} \pm 1000\,\mathrm{K}$.

6.3 Solar Heating

From (4.6) and (4.7), the luminosity of the Sun, $\mathcal{L}_\odot$, in terms of its radius, $R_\odot$, and temperature, $T_\odot$, is

$$\mathcal{L}_\odot = 4\pi R_\odot^2 \sigma T_\odot^4 (\mathrm{W/m^2}) \tag{6.14}$$

and from Chapter 4.4.4, it is also expressible in terms of the solar constant and total sphere area at the Earth's orbit, (4.13):

$$\begin{aligned} \mathcal{L}_\odot &= 4\pi r^2 \mathfrak{S} = 3.826(8) \times 10^{33}\,\mathrm{erg/s.} \\ &= 3.826(8) \times 10^{26}\,\mathrm{W} \end{aligned}$$

(Allen 1973, p. 169). The flux at distance r from the Sun's center is then

$$\mathcal{F}(r) = \mathcal{L}_\odot/[4\pi r^2](\mathrm{W/m^2}) \tag{6.15}$$

The power, P, striking any cross-sectional area, α, normal to the direction of radiation (e.g., an area of a planetary surface with the Sun at the zenith) is

$$P = \alpha\mathcal{F}(\mathrm{W}) \tag{6.16}$$

and the power absorbed over that area is

$$P_a = P(1 - A) = \alpha\mathcal{F} \cdot (1 - A)(\mathrm{W}) \tag{6.17}$$

where A is the *bolometric albedo* (effectively the ratio of reflected to incident bolometric flux). This power must be reradiated, because otherwise, the energy would increase continuously with time and the temperature would rise without limit (contrary to both observation and the laws of thermodynamics!). The reradiated or emitted power for that area is analogous to the luminosity of a star:

$$P_e = \alpha\sigma T^4 \ (\mathrm{W}) \tag{6.18}$$

where T is the *equilibrium temperature*. Setting $P_a = P_e$, so that

$$\alpha\mathcal{F}(1 - A) = \alpha\sigma T^4 \tag{6.19}$$

$$T^4 = \frac{\mathcal{F}(1 - A)}{\sigma} \tag{6.20}$$

and we get the equilibrium temperature:

$$T = \left[\frac{\mathcal{F}(1 - A)}{\sigma}\right]^{1/4} = \left[\frac{\mathcal{L}_\odot(1 - A)}{4\,\pi\,\sigma\,r^2}\right]^{1/4} \tag{6.21}$$

Note that the cross-sectional area cancels out. However, the equilibrium temperature at the subsolar point (i.e., the spot on the Earth where the Sun is overhead) is much higher than the equilibrium temperature of the planet as a whole, and is higher than the equilibrium temperature at sites where the zenith distance of the Sun exceeds 0°.

Figure 6.12 demonstrates the geometry when the Sun impinges on the area through a zenith distance, z, in which case (6.17) becomes:

$$\begin{aligned} P_a &= P(1 - A) \\ &= \alpha \cos z\, \mathcal{F}(1 - A) \ (\mathrm{W}) \end{aligned} \tag{6.22}$$

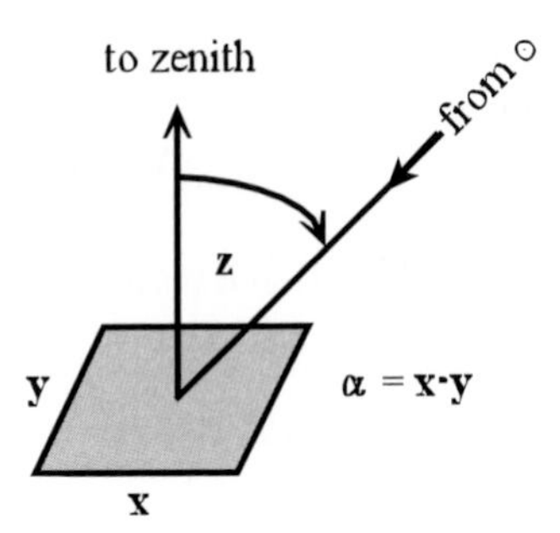

Fig. 6.12. Geometry of insolation

The factor α cos z arises because, if the area is xy, and the Sun, the x-axis, and the zenith lie in the same plane, the input flux is spread over a cross-sectional area:

$$(x \cos z)y = \alpha \cos z.$$

The emission, however, is from the total area, $\alpha = xy$.

Again, setting the absorbed and emitted power, $P_a = P_e$, equal,

$$\alpha \cos z\, \mathcal{F}(1 - A) = \alpha\sigma T^4 \tag{6.23}$$

and designating the equilibrium temperature of (6.21) as $T_{\alpha,0}$ and that of (6.23) as $T_{\alpha,z}$, we get:

$$T_{\alpha,z} = T_{\alpha,0}[\cos z]^{1/4} \tag{6.24}$$

Looked at in this way, we see that $T_{\alpha,0}$ is the maximum possible temperature achievable during the day at the subsolar point, where the Sun is overhead. At any other place on the planet the maximum temperature is found at $z = z_{\text{min}} = h_{\text{max}}$, the minimum zenith distance and maximum altitude of the Sun at local noon, when the Sun is on the observer's celestial meridian (see Chapter 2.1.4).

These temperatures are purely *local* temperatures, in the sense that every part of the planet can be thought of as having a different local temperature. If the Sun is not in the sky at all, the local temperature will decrease as the reradiated energy matches the reduced heat flux into the designated area.

We may make a similar calculation for the planet as a whole.

From the standpoint of the radiation impinging on the planet, the planet appears as a flat disk. Designating an incrementally small area, da, of this disk of radius $\Re$, we obtain for the absorbed power,

$$P_a = \int \mathcal{F}(1 - A)\, \mathrm{d}a \text{ (W)} \tag{6.25}$$

where the integration is taken over all sunlit areas. The integral can be taken over disk segments of width $\mathrm{d}\theta$ and thickness $\mathrm{d}r$, where r is the apparent distance of the area element from the subsolar point. Thus,

$$\mathrm{d}a = \int \mathrm{d}\theta \int r\,\mathrm{d}r \tag{6.26}$$

where the integration in theta is taken over 2π and the radius is taken from 0 to $\Re$, the radius of the planet. The integration over ring segments of radius r, width $\mathrm{d}r$, and angular wedge $\mathrm{d}\theta$ within the planetary disk is illustrated in Figure 6.13.

Because $\int \mathrm{d}\theta = 2\pi$ and $\int r\mathrm{d}r = \frac{1}{2}\Re^2$, and with the assumption that we can assign a mean or effective bolometric albedo to the planet as a whole, we arrive at the equation for the area of a circle of radius $\Re$ times the solar flux at the planet and the average fraction of absorption, $< 1 - A >$:

$$P_a = \pi\Re^2\mathcal{F} < 1 - A > \text{ (W)} \tag{6.27}$$

The reradiated power depends on how well the flux is distributed over the planet, i.e., it depends on the rotation as well as the atmospheric convection of the solar heat energy. Ignoring the latter, we can discuss two cases: *rapid* and *slow* rotations. An example of the latter would be a planet that is locked into its orbital angular rate so that the rotation period is equal to the revolutionary period (there is no such planet in the solar system). Such a planet would reradiate from half its surface area, whence the emitted power becomes:

$$P_\mathrm{e} = 2\pi\Re^2\sigma T^4 \text{ (W)} \tag{6.28}$$

so that

$$T_\mathrm{s} = \left[\frac{\mathcal{F} < 1 - A >}{2\sigma}\right]^{1/4} = \left[\frac{\mathcal{L}_\odot < 1 - A >}{8\pi\sigma r^2}\right]^{1/4} \tag{6.29}$$

This is then an estimate of the mean equilibrium temperature of the sunlit hemisphere of a slowly rotating planet, or at any rate a planet rotating so

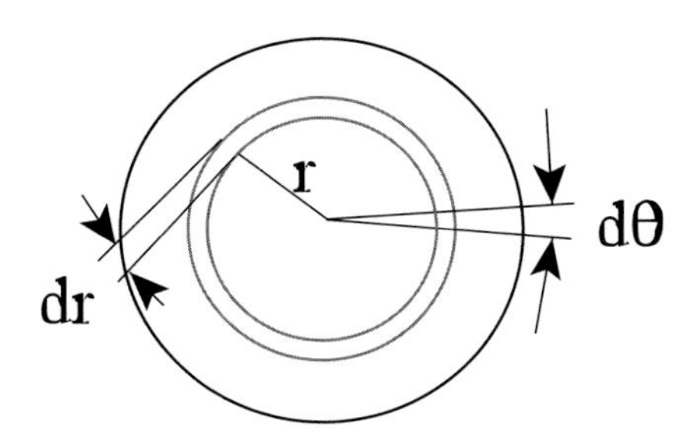

Fig. 6.13. Integration over the planetary disk, centered on the subsolar point

slowly that a relatively insignificant amount of thermal emission is coming from the night side of the planet. The detection of some thermal radiation from the night side of Mercury in the 1960s proved conclusively that Mercury was not locked in a 1:1 spin–orbit coupling with the Sun; Doppler radar mapping later showed that the lock-in rate is, rather,

$$3P_{\text{rotations}} = 2R_{\text{revolutions}}.$$

If the planet is a *rapid* rotator, on the other hand, the emitted power, from (6.18), becomes:

$$P_e = 4\pi\Re^2\sigma T^4 \text{ (W)} \tag{6.30}$$

because all of the planet now contributes to the emission, so that, on average,

$$T_{\text{r}} = \left[\frac{\mathcal{F} < 1 - A >}{4\sigma}\right]^{1/4} = \left[\frac{\mathcal{L}_{\odot} < 1 - A >}{16\pi\sigma r^2}\right]^{1/4} \tag{6.31}$$

Note that in neither (6.29) nor (6.31) is there a dependence on the radius of the planet.

The relationship between these mean equilibrium temperatures is

$$T_{\text{s}} = T_{\text{r}} \cdot 2^{1/4}. \tag{6.32}$$

This treatment has seemed to depend on whether planets act like black bodies. They do not, in fact. However, the true temperature is related to an effective temperature, i.e., the temperature the planet would have if indeed it were a black body, through the expression:

$$T_{\text{eff}} = \varepsilon\, T_{\text{true}} \tag{6.33}$$

where ε is called the *emissivity*[1]. If this quantity were known, the RHS of (6.32) could be inserted into (6.29) and (6.31), to give the true temperature in terms of mean albedo and emissivity, the distance from the Sun, and, of course, the radiation properties of the Sun.

Example 6.1

Here we compute the Earth's equilibrium temperature.

---

[1] N.B: This emissivity is not that defined in (4.22).

$\mathcal{L}_{\odot} = 3.83 \times 10^{26}$ W, $\mathcal{F} = 1362\,\mathrm{Wm}^{-2}$, $A_{\oplus} = 0.307$, and, with $\Re_{\oplus} = 6.378 \times 10^6$ m, the absorbed power is, from (6.26),

$$P_a = \pi \Re^2 \mathcal{F} < 1 - A >= 1.21 \times 10^{17}\,\mathrm{W}.$$

Because the Earth can be assumed to be a rapidly rotating planet, we have, for the emitted power, from (6.29),

$P_e = 4\pi\Re^2\sigma T^4$, and, therefore, the effective equilibrium temperature for this (rapid rotation) case, T_{r}, is

$$T_{\mathrm{r}} = \left[\frac{\mathcal{L}_{\odot} < 1 - A >}{16\,\pi\,\sigma \mathrm{r}^2}\right]^{1/4} = \left[\frac{3.83 \times 10^{26} \times 0.693}{16\,\pi\,5.67 \times 10^{-8} \cdot [1.496 \times 10^{11}]^2}\right]^{1/4}$$
$$= 254.0\,\mathrm{K}.$$

For more examples, see Schlosser et al. (1991/4, Ch. 17, 18). Lewis (1995/7, Ch. V) has a calculation for the giant planets.

Usually, equilibrium temperatures differ from actual observed temperatures for many reasons, such as thermal inertia on the sunset side of a slowly rotating planet, atmospheric circulation effects, internal heat sources and the greenhouse effect. A planetary atmosphere acts to scatter radiation from the Sun (and other astronomical objects!), which contributes to the local heating budget. Also, it is well known that H_2O and CO_2 are important *greenhouse* gases, which strongly absorb in the IR preventing complete cooling by radiation from the surface and lower atmosphere (an extremely important condition on Venus but also present on the Earth). Moreover, atmospheric molecules cool the (upper) atmosphere through the emission process by radiating into space. A source of cooling in the Earth's atmosphere at 20 km altitude, for example, is the radiation due to the $15\,\mu$m line of CO_2. These processes are dealt with in detail in Goody (1995).

The presence of oceans on Earth is thought to have played a major role in the regulation of CO_2 and the development of photosynthetic vegetation, which consumes CO_2 and produces O_2 and H_2O. The carbon dioxide-consuming plants have played a further critical role in the evolution of Earth's atmosphere into a nitrogen–oxygen one. In Milone & Wilson, (2008, Chapters 10 and 12) we discuss the structure of the atmosphere, circulation effects, and the properties of individual planets.

This concludes our general discussion of the heating and temperature of a terrestrial planet. We return next to the rocks and minerals of these planets' interiors, concentrating once more on the material of that planet we know best, the Earth.

References

Allen, C. W., Ed. 1973. *Astrophysical Quantities,* 3th Ed. (London: The Athlone Press, University of London)

Anderson, D. L. 1989. *Theory of the Earth.* (Boston: Blackwell Publications)

Bukowinski, M.S.T. 1999. "Earth Science: Taking the Core Temperature," *Nature*, **401**, 432–433.

Goody, R. 1995. *Principles of Atmospheric Physics and Chemistry* (Oxford: University Press).

Lewis, J. S. 1995/7. *Physics and Chemistry of the Solar System* (San Diego, CA: Academic Press).

Milone, E. F., and Wilson, W. J. F. 2008. *Solar System Astrophysics: Planetary Atmospheres and the Outer Solar System.* (New York: Springer)

Schlosser, W., Schmidt-Kaler, Th., and Milone, E. F. 1991/4. *Challenges of Astronomy* (New York: Springer-Verlag).

Challenges

[6.1] At what solar zenith angle is the local temperature equal to the global equilibrium temperature of a rapidly rotating planet (assuming black body radiation is solely responsible for the temperatures).

[6.2] Suppose internal heat were the only source of heat flux to the Earth's surface. What would be the equilibrium temperature of the Earth?

[6.3] Suppose the heat flux from the interior were to equal the average heat flux from the Sun, and that it was uniform around the Earth. What equilibrium temperature would result? Would this calculation need to be refined further because of the effects of a higher temperature on the Earth's surface and atmosphere?

7. Rocks and Minerals

On other terrestrial planets, we have barely scratched the surface in determining the composition and state of their rocks and minerals. Therefore, we must rely on the extensive geological knowledge of the Earth as our primary guide. We begin by making some basic distinctions concerning rocks and minerals.

7.1 Rocks

Rocks differ from minerals in that minerals are crystalline solids having a definite chemical composition (e.g., quartz (SiO_2)), and rocks are consolidated assemblages of minerals. Rocks can be classified as igneous, sedimentary, or metamorphic, as follows.

7.1.1 Igneous Rocks

Igneous rocks are those that solidified directly from molten rock, in the form of either magma or lava. *Magma* is molten rock below the surface of the Earth, whereas *lava* is magma that comes out onto the surface of the Earth.

The sizes of crystal grains in igneous rocks depend on the rate of cooling. If the rock cools quickly then there is no time for large crystals to form, and the crystals are small or absent. If the rock cools slowly then the crystals are larger.

Igneous rocks are divided into two types, based on grain size:

1. *Volcanic rocks* cooled quickly and therefore have a very fine grain or no grain structure. They are often porous (containing pores or air pockets) and are often associated with volcanic outthrows or lava flows.
2. *Plutonic rocks* formed in intrusions of magma which remained buried. They therefore cooled slowly and have a coarse grain structure and low porosity.

7.1.2 Sedimentary Rocks

Sedimentary rocks are formed after igneous and other rocks are eroded (weathered), washed away, and deposited as sediments. These sediments become overlain by other sediments, and are eventually compressed to form rock. Two examples are limestone, which is the main constituent of the Rocky Mountains, and sandstone.

7.1.3 Metamorphic Rocks

Metamorphic rocks are formed when sedimentary or igneous rocks are metamorphosed (changed) by high temperature and pressure, without melting.

The average composition of the Earth's crust is approximately 94% igneous rock, 6% sedimentary rock, and <1% metamorphic rock.

7.1.4 The Geochemical Cycle

The *geochemical cycle* is shown schematically in Figure 7.1. Heavy lines show the main cycle, in which magma solidifies to form igneous rocks that are subsequently weathered and compressed to form sedimentary rocks. These undergo metamorphism to produce metamorphic rocks, and the metamorphic rocks are remelted to form magma. Many other paths are also possible, two examples of which are shown by lighter lines.

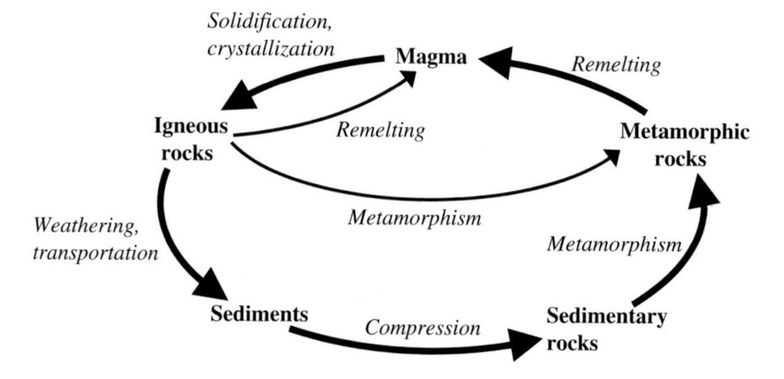

Fig. 7.1. The geochemical cycle

7.2 Minerals

Each different mineral has its own chemical composition (e.g., $MgSiO_3$) and/or crystal structure (e.g., face-centered cubic) and/or crystal size (e.g., large, small, absent), etc.

7.2.1 Crystal Structure

There are many possible structures for crystals. We will describe two of these in detail here.

We begin by packing spheres of equal size into a single layer in the densest possible manner, as shown in Figure 7.2. Three properties of this close-packing arrangement are:

1. The spheres form a three-way grid (lines 120° apart), as indicated by the lines labeled A in Figure 7.2.
2. Any three spheres in mutual contact form an equilateral triangle (e.g., the triangle labeled B in Figure 7.2).
3. Every sphere has six equidistant neighbors touching it, forming a hexagon (C in Figure 7.2).

If a second, identical, layer is stacked on top of this one, then the central planes of the two layers will be closest together when the second layer sits as low as it can over the first layer. This happens when the spheres in the upper layer lie over the holes (centers of the equilateral triangles) in the lower layer. Figure 7.3a shows the resulting arrangement. Here, solid circles represent the upper layer and dashed circles represent the lower layer. Figure 7.3b shows a more open view (with a slightly expanded scale) which may make the

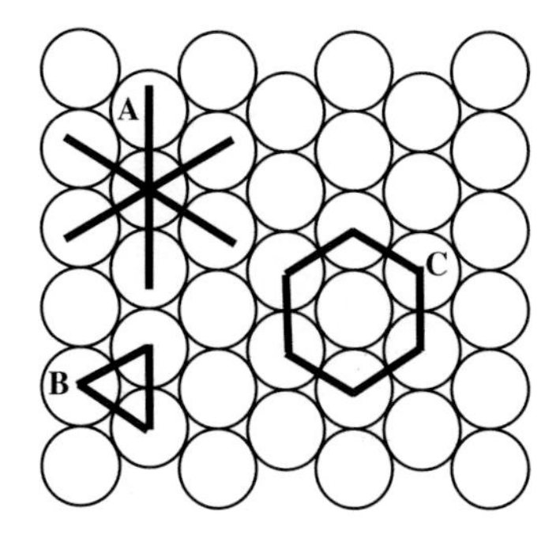

Fig. 7.2. A single layer of closely packed spheres

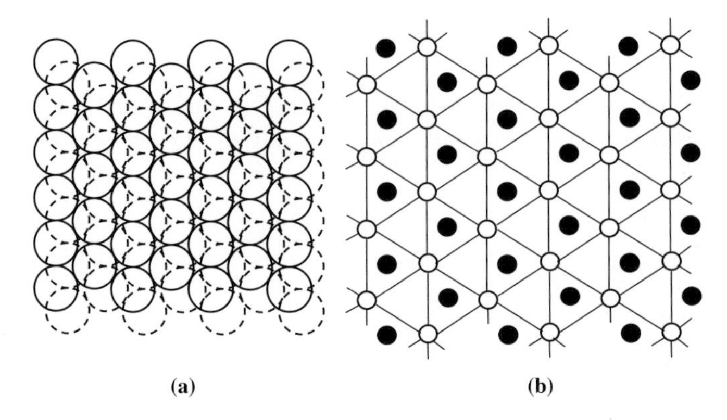

Fig. 7.3. Two layers of closely packed spheres with the upper layer resting as low as possible over the lower layer (see text)

arrangement clearer. Here, only the centers of the spheres are shown. Open circles indicate spheres in the upper layer, and filled circles indicate spheres in the lower layer.

A careful examination of Figure 7.3 shows that some of the holes in the upper layer lie over *spheres* in the lower layer and others lie over *holes* in the lower layer. The fact that there are two types of holes in the upper layer means that there are now two different ways to stack a third (identical) layer on top of this one.

1. Stack the third layer so that its spheres are directly over the *spheres* in the layer two below it. If we continue stacking in this fashion, then the layers in Figure 7.3 will alternate in an xyxyxy... pattern. The solid circles on the left side of Figure 7.3 then represent ions in layers 1, 3, 5, ... and the dashed circles represent ions in layers 2, 4, 6,....

   Figure 7.4 shows a portion of Figure 7.3. Consider the hexagon of solid circles in Figure 7.4 alternating with the triangle of dashed circles. The

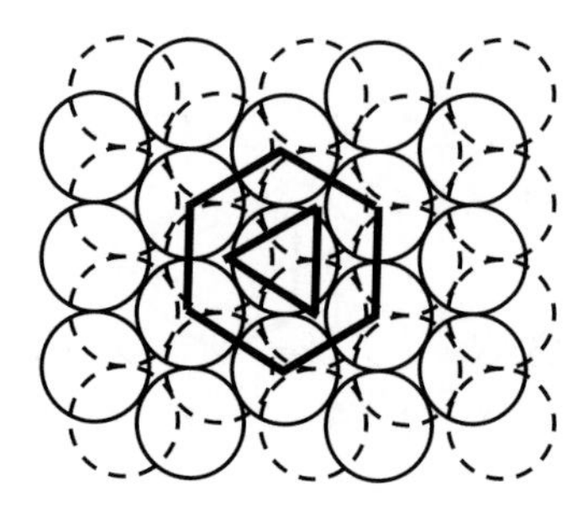

Fig. 7.4. A small section of Figure 7.3a

triangles are, of course, also parts of hexagons. Figure 7.5 shows (on the left) a three-dimensional view of four successive layers of this arrangement and (on the right) an exploded view of the same four layers.

This method of stacking layers of spheres is called *hexagonal close packing* (HCP).

2. Stack the third layer so its spheres are over *holes* in *both* of the two layers below it. If we continue stacking in this fashion, then the spheres in any given layer are over the spheres in the layer *three* below, and the pattern of stacking that results is xyzxyzxyz....

   Figure 7.6 shows a view through such a crystal (compare with the right-hand diagram in Figure 7.3).

   As we look down through the crystal, open circles represent layers 1, 4, 7, ..., filled circles represent layers 2, 5, 8, ..., and stippled circles represent

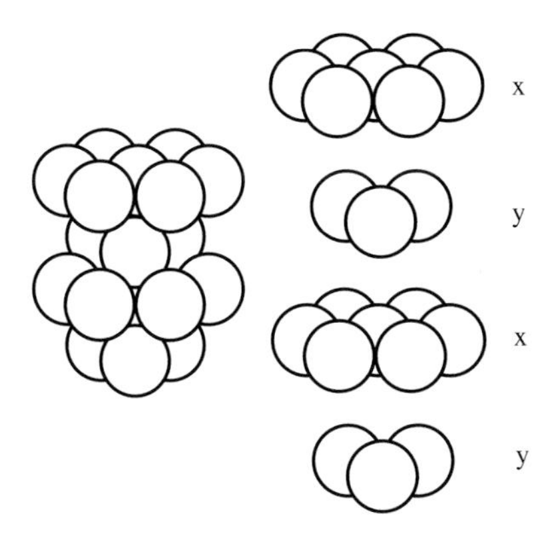

Fig. 7.5. Perspective (*left*) and exploded (*right*) views of four consecutive layers in hexagonal close packing (HCP)

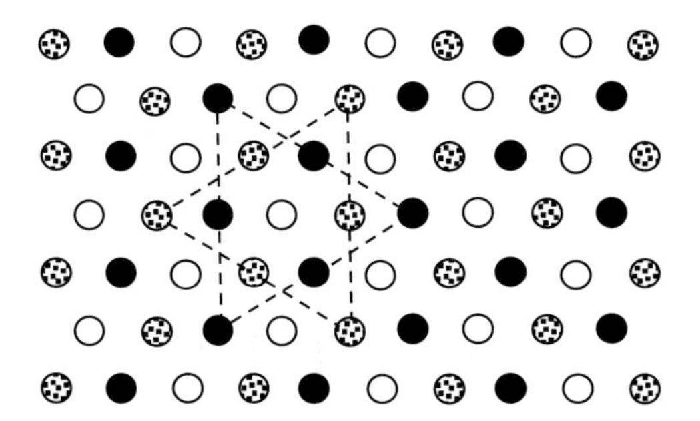

Fig. 7.6. A view through a crystal in which the layers are stacked xyzxyzxyz..., as explained in the text

layers 3, 6, 9, ... The star-shaped arrangement shown in Figure 7.6 is seen again in Figure 7.7.

Other arrangements are also possible, such as xyzyxyzyx..., but these do not concern us here.

Figure 7.7a shows the star-shaped portion of Figure 7.6. Figure 7.7b shows an exploded three-dimensional view of four successive layers in this portion. There are a total of 14 ions in these four layers. Figure 7.7c shows these same four layers pushed together as they exist in the crystal; and Figure 7.7d shows schematically the locations of the 14 ions involved. Figures 7.7c and d have been rotated to the right and away from the viewer compared to Figure 7.7b, as indicated by the dashed line (which is the same line in all three diagrams).

Figures 7.7c and d show that the 14 ions make up a cube. Eight ions form the corners of the cube, and the other six ions are located at the centers of the six faces. This xyzxyz ... pattern is therefore referred to as *cubic close packing* (CCP), and forms a *face-centered cubic* (FCC) lattice.

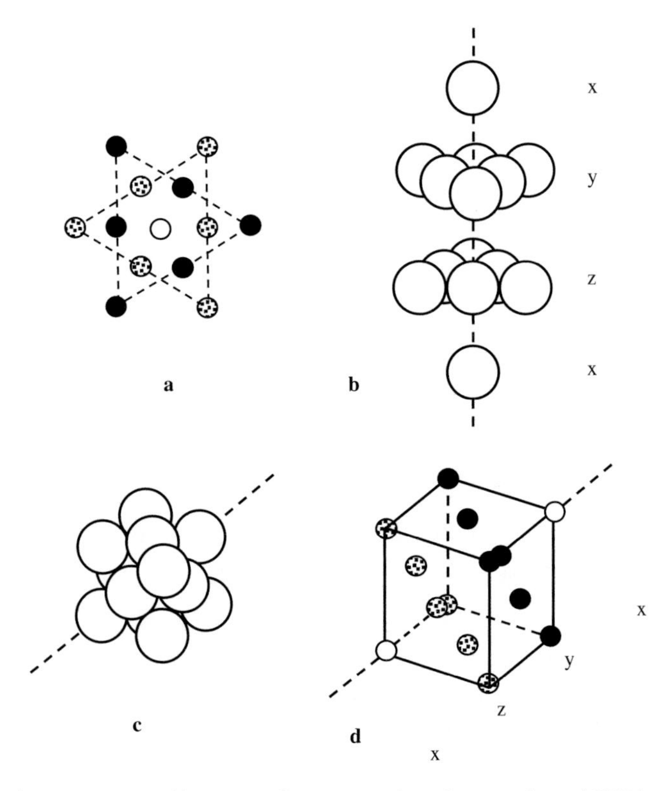

Fig. 7.7. Arrangement of layers and ions in cubic close packing (CCP), producing a face-centered cubic (FCC) crystal

Note that the layering formed by the top and bottom faces and the face-centered ions in the FCC pattern (Figures 7.7c and d) is not the same as the layering we used to generate the pattern (Figures 7.7a and b). The relationship between the two is shown by the arrangement of open, filled, and stippled circles in Figure 7.7d, which correspond to the same symbols in the upper left-hand diagram, and in Figure 7.6.

7.2.2 Crystal Density

If we pack steel balls in a box, then the balls in each layer are not directly affected by the balls in the layer two below. As a result, the layers are the same distance apart whichever arrangement (xyxyxy or xyzxyz) is used, and the overall density (g/cm^3) is the same for both HCP and CCP (or FCC).

However, ions of the same sign (+ or −) repel each other even if they are not physically touching. Therefore, an ion in a given layer is repelled more by the layer two below if the ion in the upper layer is over an ion in the lower layer (xyxyxy) than if it is over a hole in the lower layer (xyzxyz). As a result, a CCP (FCC) crystal has a greater density in kg/m^3 than an HCP crystal made of the same ions.

7.2.3 Interstitial Holes

When one layer is stacked on top of another to form a pair of layers in contact, holes (spaces) are created between the ions in one layer and the ions in the other layer. They are not the same as the "two-dimensional" holes discussed above, which are at the centers of the equilateral triangles *in* each layer. If we think of the midplanes of the layers as being located at $x = 0$ and $x = 1$, then for the close-packed layers discussed here the holes will be near $x = 1/2$. These holes between two layers are referred to as *interstitial holes.*

Interstitial holes have a three-dimensional structure, in that each hole is surrounded by a small, three-dimensional framework of ions. There are two types of interstitial holes, depending on the number of ions (shape of the framework) surrounding the hole:

7.2.3.1 Tetrahedral Holes If an equilateral triangle of ions in the lower layer has an ion above it, then the hole between the two layers is surrounded by four ions, all in mutual contact. The hole is then at the center of a *tetrahedron*: a four-sided structure, each face of which is an equilateral triangle.

Two examples of tetrahedral holes are shown by heavy lines in Figure 7.8a, with a three-dimensional view in Figure 7.8b. (Remember that open and filled circles represent only the centers of the ions; the ions themselves are larger

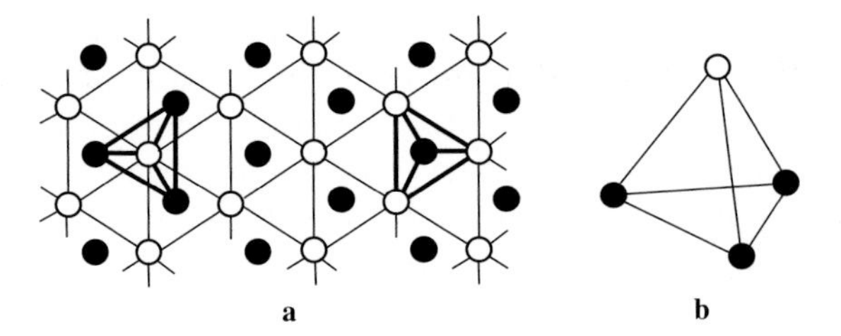

Fig. 7.8. Tetrahedral holes between adjacent layers of a crystal. *Open circles* represent ions in the upper layer and *filled circles* represent ions in the lower layer

and in mutual contact.) The left-hand tetrahedron in Figure 7.8a points out of the page, and the right-hand one points into the page.

7.2.3.2 Octahedral Holes If the equilateral triangle in the lower layer does *not* have an ion over it in the layer above, then the hole between the layers is surrounded by *six* ions.

Two examples are shown in Figure 7.9. In the left-hand example, the ions surrounding the hole are labeled from 1 to 6. The other example has been shaded to bring out the three-dimensional structure of the ions surrounding the hole more clearly.

Two of the ions in the lower layer (1 and 5) form a square with two in the upper layer (2 and 6). One other ion in the upper layer (3) and one in the lower layer (4) lie "above" and "below" the midpoint of the square, along its central axis.

As a result, the hole is at the center of an octahedron, an eight-sided figure made of two four-sided pyramids base-to-base.

7.2.4 The Silicate Tetrahedron

The crystals of silicate minerals (described below) can be regarded as stacked layers of oxygen ions, with silicon ions occupying a fraction of the tetrahedral

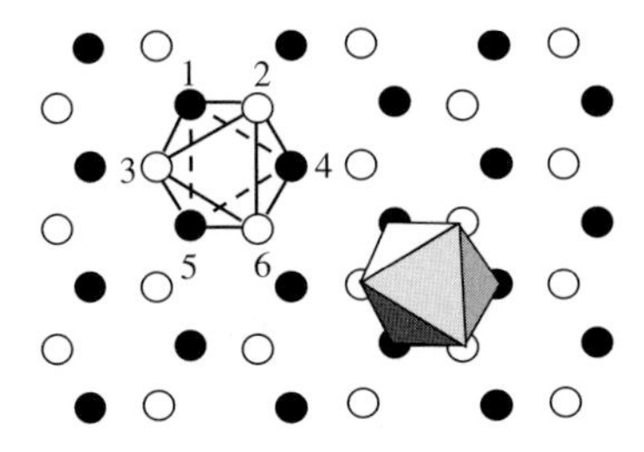

Fig. 7.9. Octahedral holes between adjacent layers of a crystal. *Open circles* represent ions in the upper layer and *filled circles* represent ions in the lower layer

sites and other metal ions occupying a fraction of the octahedral sites. The respective fractions of tetrahedral and octahedral sites occupied are different for different minerals.

The silicate tetrahedron (SiO_4) is shown in Figure 7.10, and is the basic building block of almost all silicate minerals. (In the case of stishovite and perovskite, discussed below, the basic building block is a silicate octahedron.)

The valence of silicon is +4 and that of oxygen is −2, so the valence of SiO_4 is $4 + 4(-2) = -4$. Each SiO_4 therefore bonds easily with other ions, usually those of silicon or metals.

7.2.5 Mineral Names

Mineral names are somewhat arbitrary, because:

1. Minerals with the same crystal structure but different chemical compositions can have the same name. This happens because certain types of ions (e.g., magnesium and iron) are so similar to each other in size and chemical properties that one can replace the other in the octahedral sites with no change to the crystal structure.

   An example is olivine, a silicate mineral in which each occupied octahedral site contains an ion of either magnesium (Mg) or iron (Fe). If all such sites are occupied by Mg, then the mineral is called *forsterite* (Mg_2SiO_4), and if all are occupied by iron then the mineral is called *fayalite* (Fe_2SiO_4).

   However, these are simply end-members of a continuous series of possible compositions, and the chemical symbol for olivine is often given as $(Mg_{(1-x)}Fe_x)_2\,SiO_4$, where x can have any value from 0 to 1. For example, $(Mg_{0.9}Fe_{0.1})_2SiO_4$ has 90% of the sites occupied by magnesium and 10% by iron. All these possible compositions are included in the name olivine.

   Olivine is described more fully in Section 7.2.8.4.

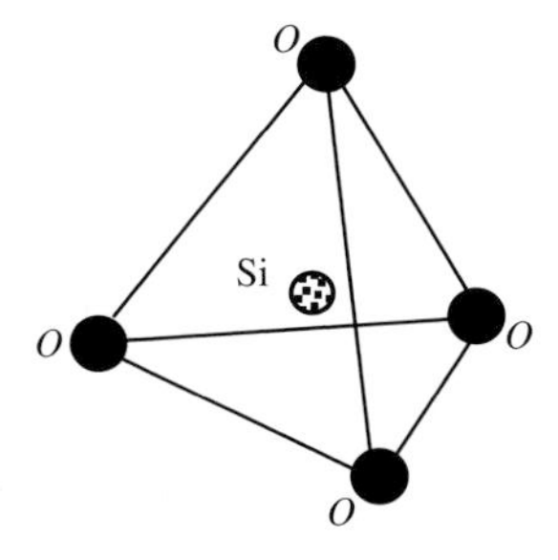

Fig. 7.10. Silicate tetrahedron. The silicon ion (*stippled circle*) occupies the tetrahedral hole at the center of the tetrahedral framework of four oxygen ions (*filled circles*)

2. Minerals with the *same* chemical composition can exist in different *phases* (different crystal lattices) and these can have different names.

   Olivine and silicate spinel (hereafter referred to simply as "spinel") provide an example. Both have the chemical symbol Mg_2SiO_4 (or Fe_2SiO_4, etc), but they differ in crystal structure: in olivine the oxygen ions are arranged in an HCP lattice and in spinel they are arranged in a CCP (or FCC) lattice.

   Spinel is a high-pressure phase of olivine. Olivine is common in the Earth's upper mantle, but deeper in the mantle where pressures are higher it undergoes a *phase transition* to the more closely packed crystal structure of spinel. Spinel therefore has the same chemical symbol as olivine, but is about 10% denser. Spinel is discussed again in Sections 7.2.8.4 and 7.4.2.

   A more familiar example of a substance changing name when it undergoes a phase transition is, of course, water changing to ice.

3. Some minerals have different names even though they have the same chemical composition *and* crystal structure, because they differ in crystal size or some other characteristic.

7.2.6 Composition of the Earth's Crust

The mean chemical composition of the Earth's crust, excluding some trace constituents, is given in Table 7.1.

That is, oxygen atoms make up 60.5% of all atoms in the crust, but, because oxygen atoms are lighter than all of the others listed except hydrogen, they make up only 46.60% of the weight of the crust.

Oxygen and silicon together make up 74% of the weight of the crust, or 81% of the mass. It is therefore clear that the most common minerals will be those involving oxygen and silicon: *the silicates.*

Table 7.1. Mean composition of the Earth's crust

Element	Symbol	Weight %	Atom %
Oxygen	O	46.60	60.5
Silicon	Si	27.72	20.5
Aluminum	Al	8.13	6.2
Iron	Fe	5.00	1.9
Calcium	Ca	3.63	1.9
Sodium	Na	2.83	2.5
Potassium	K	2.59	1.8
Magnesium	Mg	2.09	1.4
Titanium	Ti	0.44	
Hydrogen	H	0.14	
Phosphorus	P	0.12	
		99.29	

7.2.7 Oxygen-to-Silicon Ratio in Chemical Symbols

The basic crystal unit of most silicate minerals is the SiO_4 tetrahedron. If the tetrahedra are isolated from each other (i.e., they are not linked by sharing oxygen atoms), then the mineral will contain four oxygen atoms for every silicon atom and the chemical symbol will contain SiO_4, e.g., olivine, Mg_2SiO_4.

However, if some oxygen atoms are shared between adjacent tetrahedra (i.e., the tetrahedra are linked), then on average there are less oxygen atoms per silicon atom in the mineral. This sharing changes the chemical symbol, even though the basic crystal unit is still the SiO_4 tetrahedron; e.g., pyroxene, $MgSiO_3$; quartz, SiO_2.

7.2.8 Oxygen-to-Metal Ratio and the Classification of Silicate Minerals

One way to classify silicate minerals is by their oxygen-to-metal ratio.

The most common metals in silicate minerals are silicon, magnesium, iron, calcium, potassium, and sodium, but others are also possible.

The most common ratios are, in order of decreasing oxygen abundance (or increasing metal abundance):

$$\text{oxygen:metal} = 2{:}1\ 8{:}5\ 3{:}2 \text{ and } 4{:}3$$

7.2.8.1 2:1 Ratio: Silica (SiO_2) Silica has different mineral names, depending on the crystal structure. Two examples are quartz and stishovite.

Quartz is a framework silicate: not all oxygen ions in the close-packing arrangements discussed above are actually present, and the remaining tetrahedra form a framework.

Figure 7.11 shows a single layer of linked tetrahedra in a quartz crystal. Large circles represent oxygen ions, and small circles represent silicon ions (dashed if hidden below an oxygen ion). The bases of the tetrahedra share oxygen ions to form a repeating pattern of hexagons, while the tetrahedra themselves point alternately up and down around each hexagon.

The upward-pointing tetrahedra in this layer share their oxygen ions with the downward-pointing tetrahedra in the next layer.

The SiO_4 tetrahedra therefore link to form a three-dimensional framework in which each tetrahedron shares all four oxygen ions with its neighbors. The result is that, on average, there are two oxygen ions for every silicon ion, and the chemical symbol for silica is SiO_2, even though it is made up of linked SiO_4 tetrahedra (or linked SiO_6 octahedra).

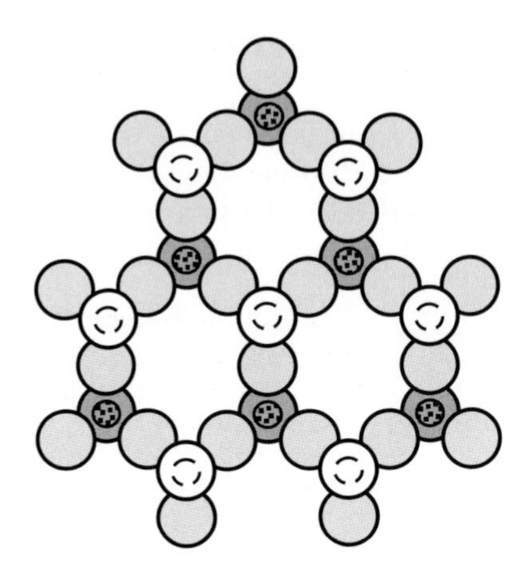

Fig. 7.11. A single layer of linked tetrahedra in a quartz crystal

Stishovite is a high-pressure (∼200 kbar) phase of quartz, made up of SiO_6 octahedra. The oxygen ions are shared between adjacent octahedra so that on average there are again two oxygen ions for every silicon ion, and the chemical symbol is still SiO_2.

7.2.8.2 8:5 Ratio: Feldspar Group: $M_2AlSi_2O_8$, where M_2=two metal ions. One of the metal atoms in the chemical symbol is potassium, K, sodium, Na, or calcium, Ca, and the other is aluminum, Al, or silicon, Si. Three examples of feldspars are:

$$\text{Orthoclase feldspar}: \quad KAlSi_3O_8$$

$$\text{Plagioclase feldspar}: \quad \begin{cases} NaAlSi_3O_8 \text{ (albite)} \\ CaAl_2Si_2O_8 \text{ (anorthite)} \end{cases}$$

Feldspars are framework silicates like quartz. For example, in orthoclase the Al^{3+} replaces one of every four Si^{4+} in the tetrahedral sites in quartz, and the charge imbalance is compensated by introducing K^+ ions into octahedral sites.

7.2.8.3 3:2 Ratio: Pyroxene Group: $M_2Si_2O_6$, where M_2 = two metal ions. The metal atoms are commonly Ca, Mg, or Fe, or less commonly K, Al, Ti, or Na.

Two examples of pyroxenes are enstatite, $MgSiO_3$ (=$Mg_2Si_2O_6$) and diopside, $CaMgSi_2O_6$.

As illustrated in Figure 7.12, enstatite forms one end-member of a continuous series of possible compositions with ferrosilite, $FeSiO_3$, in which iron replaces the magnesium of the enstatite. The enstatite–ferrosilite series is referred

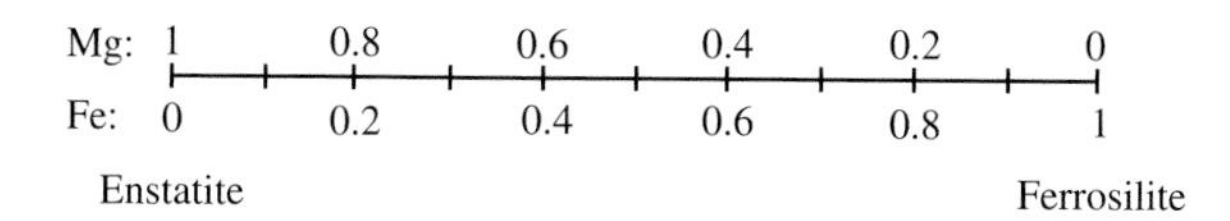

Fig. 7.12. Enstatite–ferrosilite series

to as $(Mg,Fe)SiO_3$. For example, the pyroxenes in the upper mantle have a composition near $(Mg_{0.9}Fe_{0.1})SiO_3$, which means that 90% of the available spaces are filled by magnesium ions and 10% by iron ions.

The $MgSiO_3$–$FeSiO_3$ series is referred to as *orthopyroxene*, and the $CaMgSi_2O_6$–$CaFeSi_2O_6$ series as *clinopyroxene*.

The pyroxenes are chain silicates; each tetrahedron shares an oxygen ion with the next one to form chains, as shown in Figure 7.13. Counting along a chain produces a sequence of one Si, three O, one Si, three O, etc.; so the silicate portion of the chemical symbol is SiO_3. Positive metal ions (not shown) bond adjacent chains to each other.

Phase transitions in pyroxene at high pressures can produce garnet and perovskite, described in Sections 7.4.1.2 and 7.4.1.4, respectively. These also have an oxygen:metal ratio of 3:2, and also have a range of compositions.

7.2.8.4 4:3 Ratio: Olivine and Spinel: M_2SiO_4, where M_2 = two metal ions (magnesium or iron). *Olivine* has a continuous range of compositions, referred to as $(Mg, Fe)_2SiO_4$, in which forsterite, Mg_2SiO_4, and fayalite, Fe_2SiO_4, are the two end-members. The series is illustrated in Figure 7.14.

Fig. 7.13. Pyroxene chain. *Large circles* denote oxygen ions, and *small, dashed circles* denote silicon ions (hidden below oxygen ions). Adapted from Plummer and McGeary (1988, Fig. 2.11, p. 28) and reproduced with the permission of the McGraw Hill Companies

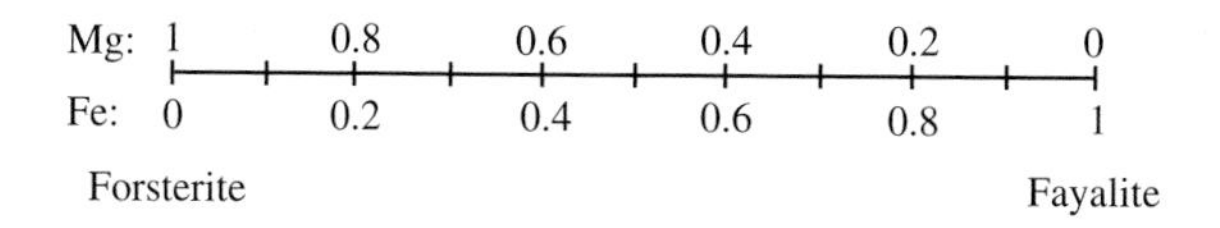

Fig. 7.14. Forsterite–fayalite series

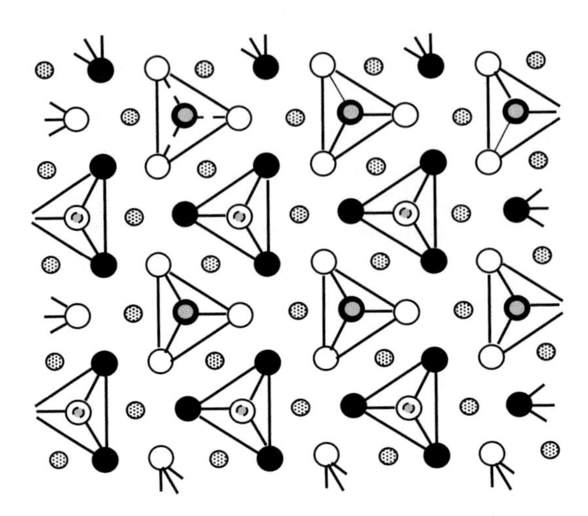

Fig. 7.15. Isolated tetrahedra of olivine. *Large open and filled circles*: oxygen ions in the upper and lower layers, respectively. *Small shaded circles*: silicon in tetrahedral sites (*dashed* if hidden below an oxygen ion). *Small stippled circles*: metal (Mg or Fe) ions in octahedral sites. Adapted from Plummer and McGeary (1988, Figure 2.10, p. 28) and reproduced with the permission of the McGraw Hill Companies

In olivine, the oxygen ions form an HCP lattice in which the SiO_4 tetrahedra are isolated from each other (that is, they do not share oxygen ions), as shown in Figure 7.15. Tetrahedra with an open circle at the center point out of the page; those with a solid circle at the center point into the page.

Silicon ions occupy the tetrahedral sites, and the metal ions (Mg or Fe) occupy the octahedral sites and bond the silicate tetrahedra together in the x, y and z directions.

Spinel is a high-pressure phase of olivine, formed at pressures >120 kbar.

Olivine and spinel are described again in Section 7.4.2.

7.3 Mineral Content of Igneous Rocks

Table 7.2 lists the mineral content of igneous rocks.

In Table 7.2 we note the following:

1. The *amphibole* group consists of chain-type silicate minerals in which every second SiO_4 tetrahedron in a chain shares an oxygen ion with a tetrahedron in an adjacent chain, forming a double-chain structure as shown in Figure 7.16. Adjacent double chains in amphibole are bonded

Table 7.2. Mineral content of igneous rocks

Density (g/cm^3)	Plutonic rock	Volcanic rock	Principal mineral content
2.6 - - - - -	Pegmatite		Quartz, feldspar
	Granite	Rhyolite	Quartz, feldspar
2.9 - - - - -	Diorite	Andesite	Feldspar, amphibole
3.2 - - - - -	Gabbro	Diabase, basalt	Feldspar, pyroxene
3.3 - - - - -			
3.5 - - - - -	Peridotite		Pyroxene, olivine
	Dunite		Olivine
	Eclogite		Pyroxene, garnet

together by positively charged ions (not shown), as is the case with the single chains of the pyroxenes.

2. *Pegmatite* is an uncommon rock that contains uranium, thorium, and other elements with unusual ion radii. These are the last to enter minerals in a solidifying magma, and therefore become concentrated in the crust.
3. *Eclogite* is obtainable by phase transitions from basalt (e.g., through subduction of basaltic crust). Eclogite may be one of the rock types making up the mantle.

Fig. 7.16. Double-chain structure of amphibole. See Figure 7.13 for an explanation of symbols. Adapted from Plummer and McGeary (1988, Fig. 2.9, p. 28) and reproduced with the permission of the McGraw Hill Companies

4. *Pyrolite* (not listed in Table 7.2) is defined as any substance which can produce basaltic magma through partial melting, leaving behind dunite and peridotite as a result. Pyrolite in this picture is regarded as a single source material for both the basalt in the crust and the dunite and peridotite in the upper mantle, and is therefore itself an important constituent of the upper mantle.

   Pyrolite is a "fictitious" rock, in that it is defined in terms of what it does, leaving its actual mineral content to be determined. It seems to be composed of 2/3 olivine and 1/3 pyroxene, plus perhaps garnet.

5. *Serpentenite* (not listed in Table 7.2) is composed mainly of olivine and pyroxene combined with water. The water lowers the melting point significantly. Serpentinite is mentioned here only to show that the presence of water complicates the description of melting or crystallization of rocks in the upper mantle.

7.3.1 Classification of Rocks by Mineral Content

Rocks may be classified according to their feldspar, silica, magnesium, and iron content.

Granite and rhyolite are silica-rich, as shown by the fact that quartz is one of the principal minerals in each one (see Table 7.2). They are also rich in feldspar, so they have a relatively high content of potassium and sodium. Such rocks are classified as *felsic* because of their high *fel*dspar and *s*ilica content. They are also more *acidic* than basalt or dunite.

Gabbro and basalt contain less silica and more magnesium, iron, and calcium than felsic rocks, and are classified as *mafic* rocks, from the words *ma*gnesium and *f*erric. They are *basic*, rather than acidic.

Rocks such as dunite and eclogite, which have an even higher magnesium and iron content and are even more basic, are often referred to as *ultramafic* or *ultrabasic.*

These classifications are shown in Table 7.3.

Table 7.3. Felsic/mafic classification

Plutonic rock	Volcanic rock	Felsic/mafic	Acidity
Granite	Rhyolite	Felsic	Acidic
Diorite	Andesite	Intermediate	
Gabbro	Diabase, basalt	Mafic	Basic
Dunite, eclogite		Ultramafic	Ultrabasic

7.4 Phase Transitions

A phase of a mineral is a particular crystal structure for that mineral.

As pressure increases with depth into the mantle, minerals undergo phase transitions in which the atoms (ions) are rearranged into a different crystal structure; for example:

$$Mg_2SiO_4 \text{ olivine} \rightarrow Mg_2SiO_4 \text{ spinel},$$

where the HCP lattice of oxygen ions in the olivine phase is transformed to the denser CCP lattice in the spinel phase.

This rearrangement can also involve one crystal structure separating into two crystal structures (i.e., the original mineral separates into two minerals) or two crystal structures merging into one (two minerals merging into a single mineral); for example,

$$Mg_2SiO_4 \text{ spinel} \rightarrow MgSiO_3 \text{ perovskite} + MgO \text{ periclase}$$

MgO has the same crystal structure as NaCl (table salt), as described in Section 7.4.1.4.

7.4.1 Phase Transitions of Pyroxene, $MgSiO_3$

7.4.1.1 Pyroxene This is the low-pressure phase, found on the Earth's surface and in the crust, e.g., enstatite, $MgSiO_3$.

The crystal structure of enstatite is illustrated in Figure 7.13 and consists of chains of corner-sharing silicate tetrahedra. Because of the sharing of oxygen ions there are, on average, three oxygen ions (valence -2) for each silicon ion (valence $+4$), so the valence of each SiO_3 unit (one silicon ion plus the base of a tetrahedron) is $3 \times (-2) + 4 = -2$. The valence of Mg and Fe is $+2$, so on average there is one Mg or Fe ion for each SiO_3 unit, giving the chemical symbol $(Mg,Fe)SiO_3$.

As described in Section 7.2.8.3, pyroxene has a variety of compositions. For example, $MgSiO_3$ enstatite forms a continuous series with $FeSiO_3$ ferrosilite by progressive replacement of Mg ions with Fe ions; or every second Mg ion in $MgSiO_3$ enstatite can be replaced by a calcium (Ca) ion, forming $CaMgSi_2O_6$ diopside.

7.4.1.2 Garnet Pyroxene undergoes a phase transition to silicate garnet, $Mg_4Si_4O_{12}$, at about 170–190 kbar pressure and $T > 2000$ K, as shown in Figure 7.20.

During the phase transition, the silicate tetrahedra change from corner-sharing chains (pyroxene, silicate unit = SiO_3) to isolated tetrahedra (garnet, silicate unit = SiO_4). In the process, one silicon ion is "released" from a tetrahedron, producing five metal ions for every three silicate tetrahedra in garnet:

$$4(SiO_3 + Mg) \rightarrow 3(SiO_4) + Si + 4Mg$$

That is, in $Mg_4Si_4O_{12}$ three of the silicon ions are inside SiO_4 tetrahedra ($3\ SiO_4 = Si_3O_{12}$), and the other silicon and the four magnesium make up the five metal ions occupying spaces between the tetrahedra.

There is a similar range of compositions for garnet as for pyroxene. Also, pyroxene in the Earth's crust and mantle is often mixed with Al_2O_3 (corundum). One Mg^{2+} ion in the garnet can then be replaced by Al^{3+} from the corundum. This produces a charge imbalance which can only be compensated if one Si^{4+} ion is also replaced by Al^{3+}, giving aluminous garnet, $Mg_3Al_2Si_3O_{12}$ (pyrope):

$$Mg_4Si_4O_{12} + Al_2O_3 \rightarrow Mg_3Al_2Si_3O_{12} + MgSiO_3$$
$$\text{(silicate garnet + corundum} \rightarrow \text{pyrope + pyroxene)}$$

At the pressures and temperatures found in the upper mantle, aluminous garnet (pyrope) is also soluble in silicate garnet, producing aluminous silicate garnet (majorite, first synthesized in the lab by Ringwood and Major (1971)).

The crystal structure of corundum consists of alternating layers of oxygen and aluminum. The oxygen forms an HCP lattice, and the aluminum occupies distorted octahedral sites between the layers of oxygen.

7.4.1.3 Silicate Ilmenite Above 200 kbar, $Mg_4Si_4O_{12}$ garnet undergoes a phase transition to a form of $MgSiO_3$ with the same crystal structure as $FeTiO_3$ ilmenite. This form of $MgSiO_3$ is generally referred to as silicate ilmenite. The crystal structure is the same as corundum (Al_2O_3), described above, with the Mg^{2+} and Si^{4+} replacing the Al^{3+} in an ordered alternate fashion.

Silicate ilmenite exhibits the same continuous sequence of compositions from $MgSiO_3$ to $FeSiO_3$ as do pyroxene and garnet.

7.4.1.4 Perovskite Above 230 kbar, $MgSiO_3$ ilmenite undergoes a phase transition to a perovskite structure, $MgSiO_3$.

Figure 7.17 shows a basic structural unit of a perovskite crystal. The Mg^{+2} (stippled) and O^{-4} (shaded) ions form a face-centered cubic lattice with the Mg at the corners and the O at the centers of the faces. The Si^{+4} ion (black) is located at the center of the cube. Figure 7.17 shows that the oxygen ions form an octahedron with the silicon at its center.

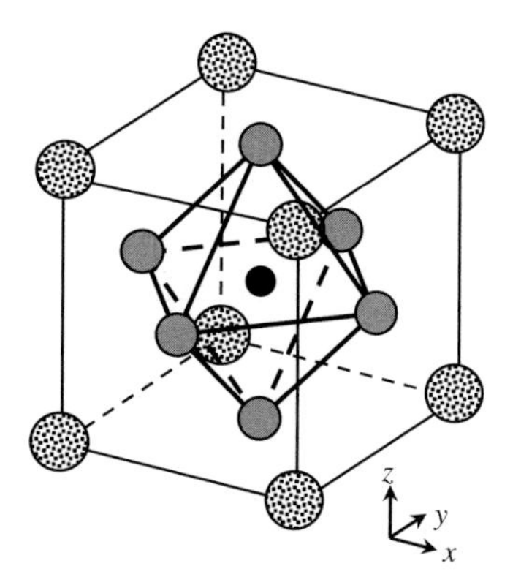

Fig. 7.17. Basic structural unit of a perovskite crystal. *Stippled circles*: Mg; *shaded circles*: O; *black circle*: Si

When we abut other cubes against this one, they all share adjoining faces and corners. Consequently, each O ion is shared between two cubes and each Mg ion is shared between eight cubes. The perovskite crystal thus consists of linked SiO_6 octahedra in which each octahedron shares all six corner oxygen ions with its neighboring octahedra, and the magnesium ions occupy the spaces between the SiO_6 octahedra.

Now imagine an infinitely repeating sequence of such cubes in all directions to form a perovskite crystal. Some thought will show that every octahedron has a Mg ion to its lower right (as viewed in Figure 7.17), and that this statement accounts for all the Mg ions; thus there is one Mg ion for every SiO_6 octahedron in the crystal. Also, if we imagine walking along each of the x, y, and z axes indicated in Figure 7.17, then starting from any silicon ion, along each axis we have one Si, one O, one Si, one O, etc. There are three axes, so there are three O ions for each Si in the crystal. The chemical symbol for silicate perovskite is thus $MgSiO_3$, despite the fact that the basic building block is an SiO_6 octahedron.

In similar fashion to olivine and pyroxene, perovskite forms a continuous series in which some Mg ions are replaced by Fe, giving $(Mg_{(1-x)}Fe_x)SiO_3$. This form of perovskite is probably the most abundant mineral in the entire Earth; it may make up more than 80% by volume of the lower mantle.

Other minerals also have phases with perovskite structure at high pressure, such as $CaSiO_3$ and $CaMgSi_2O_6$ (diopside).

However, pure $FeSiO_3$ has no perovskite phase, because at high pressure it decomposes into stishovite and wüstite:

$$FeSiO_3 \rightarrow SiO_2 \text{ (stishovite)} + FeO\text{(wüstite)}.$$

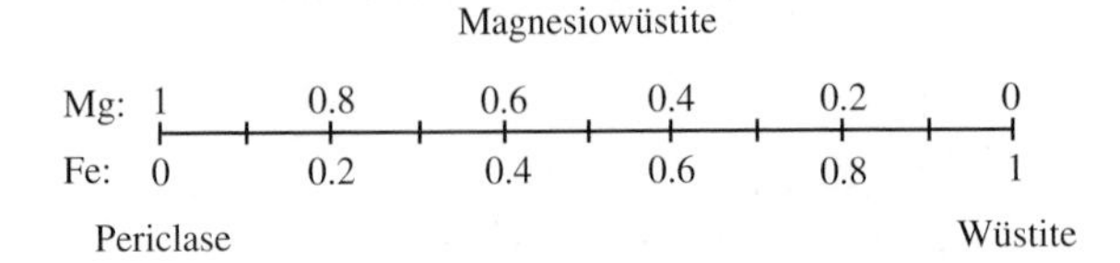

Fig. 7.18. Periclase–wüstite series

FeO (wüstite) and MgO (periclase) are two end-members of another continuous series, (Mg,Fe)O (magnesiowüstite), as indicated in Figure 7.18.

Magnesiowüstite has the same crystal structure as table salt (NaCl). The larger, negative oxygen ions form a face-centered cubic close-packed lattice, as shown in Figure 7.19, with the smaller, positive Mg or Fe ions occupying the spaces.

7.4.1.5 Phase Diagram for Pyroxene Figure 7.20 shows a pressure–temperature phase diagram for pure $MgSiO_3$ for pressures and temperatures found in the Earth's mantle.

As an example of phase transitions, the dashed, horizontal line in Figure 7.20 illustrates the phase transitions from pyroxene to garnet to ilmenite to perovskite that would occur if pressure were to increase at a constant temperature of 2000°C.

7.4.2 Phase transitions of Olivine, Mg_2SiO_4

Figure 7.21 is an isothermal phase diagram of pressure vs. iron fraction in the (Mg,Fe) part of olivine from 0% (pure Mg_2SiO_4, forsterite) to 100% (pure Fe_2SiO_4, fayalite), at a constant temperature of 1000°C.

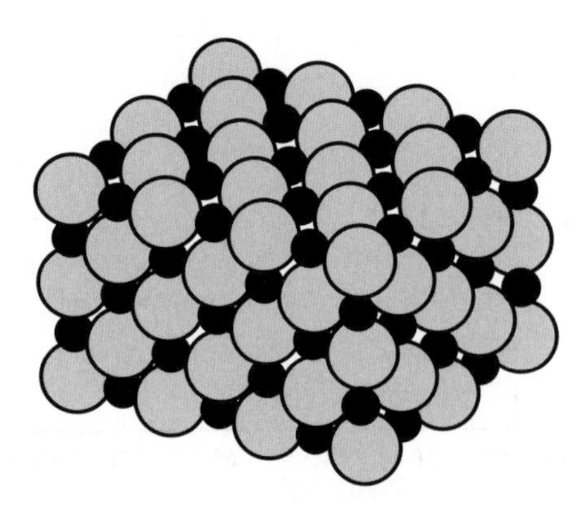

Fig. 7.19. Crystal structure of magnesiowüstite. *Large circles*: negative oxygen ions; *small circles*: positive Mg or Fe ions. Adapted from Plummer and McGeary (1988, Fig. 2.17, p. 36) and reproduced with the permission of the McGraw Hill Companies

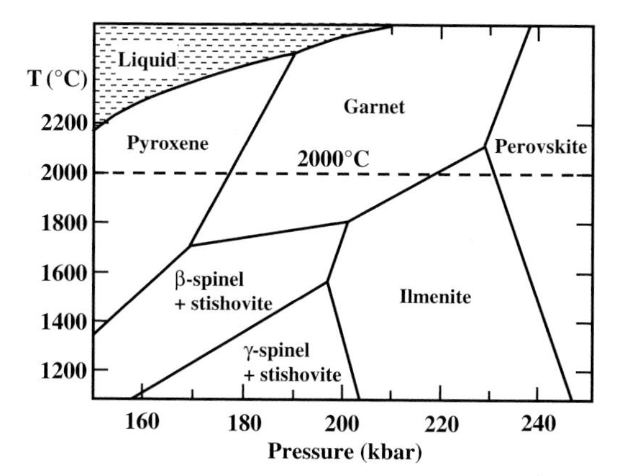

Fig. 7.20. Pressure–temperature phase diagram for $MgSiO_3$. Adapted from Anderson (1989, Fig. 16.4, p. 346), and appearing here with permission

Olivine (also called α-phase) is the low-pressure phase. The crystal structure of olivine is shown in Figure 7.15 and consists of isolated SiO_4 tetrahedra (not sharing oxygen atoms), with Mg or Fe between the tetrahedra. The oxygen ions form an HCP lattice.

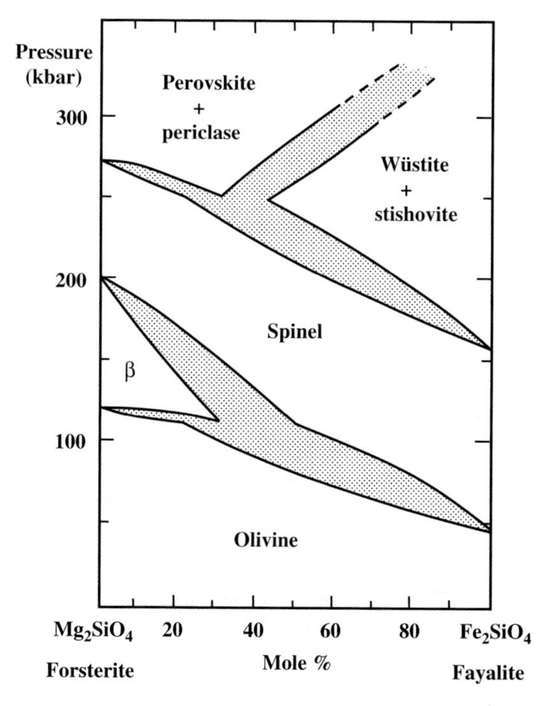

Fig. 7.21. Isothermal phase diagram at 1000°C for olivine (forsterite–fayalite system). From Bassett (1979, Fig. 6, p. 372), and reproduced here with permission

If the olivine is magnesium-rich, it undergoes a phase transition to β-phase (also called modified spinel) as pressure increases above about 200 kbar. Iron-rich olivine, however, has no β-phase and transforms directly into spinel.

β-phase is denser than olivine, with a CCP (FCC) lattice of oxygen ions. The tetrahedra are in pairs that share an oxygen atom at one corner.

β-phase (if magnesium-rich) or olivine (if iron-rich) transforms to a spinel structure at higher pressures. Spinel, also called γ-phase, has a CCP (FCC) lattice of oxygen ions with isolated tetrahedra, and is denser than either olivine or β-phase.

Many oxides crystallize with spinel structure, such $MgAl_2SiO_4$.

At higher pressures and temperatures, spinel decomposes into other minerals as shown in Figure 7.21. It is interesting that, for the expected iron-to-magnesium ratio of about 90:10 in the mantle, both pyroxene and olivine produce perovskite at lower-mantle pressures.

7.5 Densities of Minerals

Table 7.4 lists the uncompressed densities (at 1 atmosphere pressure) of various minerals mentioned above.

7.6 Seismic Discontinuities in the Earth's Mantle

Figure 5.21 shows an empirical model of P- and S-wave speeds in the Earth's mantle, and the corresponding density profile, from Dziewonski and Anderson (1981). The three significant discontinuities in wave speed at depths of 220, 400, and 670 km below the Earth's surface all correspond to abrupt changes in density, and are believed to arise from phase changes in the solid rock with increasing depth (and therefore increasing pressure) in the mantle, as described below. The Mohorovičić discontinuity, or "Moho," marks the

Table 7.4. Uncompressed densities of minerals (at 1 atm pressure) in kg/m^3

Olivine (Mg_2SiO_4) series			Pyroxene ($MgSiO_3$) series		
Forsterite	Mg_2SiO_4	3210	Enstatite	$MgSiO_3$	3200
β-phase	Mg_2SiO_4	3470	Garnet	$Mg_3Al_2Si_3O_{12}$	3560
Spinel	Mg_2SiO_4	3560	Ilmenite	$MgSiO_3$	3800
Periclase	MgO	3580	Perovskite	$MgSiO_3$	4110
Stishovite	SiO_2	4290			

crust–mantle boundary and is believed to arise from a composition change from basaltic material in the crust to dunitic or peridotitic material in the upper mantle, rather than a phase change. Variations of depth or radius with latitude and longitude are not shown; e.g., the depth of the Moho varies from 5 km below oceanic crust to 35 km below continents and even 60 km or more below mountain ranges.

7.7 Relationship of Phase Diagrams to Seismic Discontinuities

A model for the Earth's mantle is shown in Figure 7.22 (Liu 1979), with the phase transitions described above plotted on a pressure scale (top horizontal axis) and also converted to depth into the Earth (bottom horizontal axis).

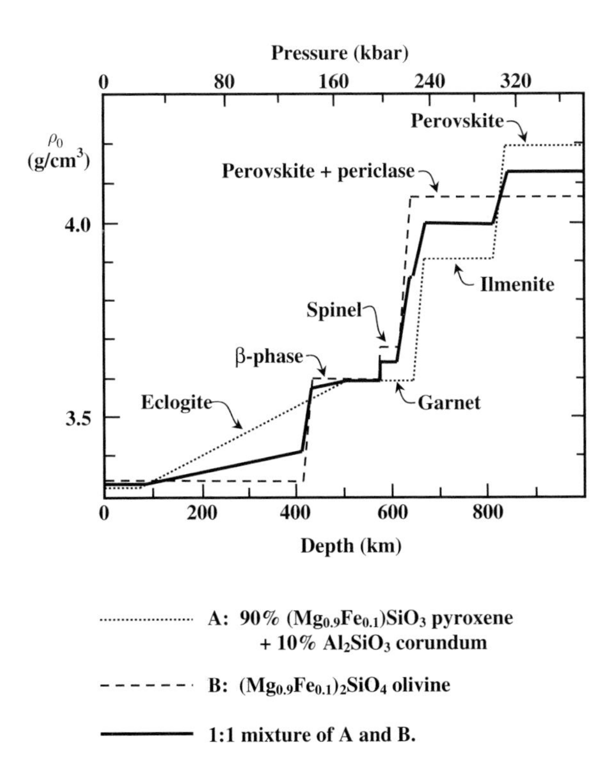

Fig. 7.22. Model of phase transitions in the mantle, with olivine, pyroxene, and corundum in the proportions given in the legend. From Liu (1979, Fig. 7, p. 195), with permission

The density, ρ_o, plotted along the vertical axis in Figure 7.22 is the uncompressed density, as in Table 7.4, not the actual density under pressure.

In this model, the 400 and 670 km seismic discontinuities are seen as arising from mineral phase transitions:

400 km discontinuity: olivine to β-phase

670 km discontinuity:

spinel to perovskite + periclase in the olivine component

and

garnet to ilmenite in the pyroxene/corundum component.

A small seismic discontinuity has sometimes been suspected between 400 and 650 km depth, and may result from the β-phase to spinel transition in the olivine component.

Another small discontinuity below 670 km has also been suspected, and may be due to the transition from ilmenite to perovskite in the pyroxene/corundum component.

7.8 The Core–Mantle Boundary and the D″ Layer

The D″ layer, shown schematically in Figure 7.23, is a region about 200–300 km thick observed at the base of the lower mantle, in contact with the core, where seismic velocities show large lateral (side-to-side, as opposed to

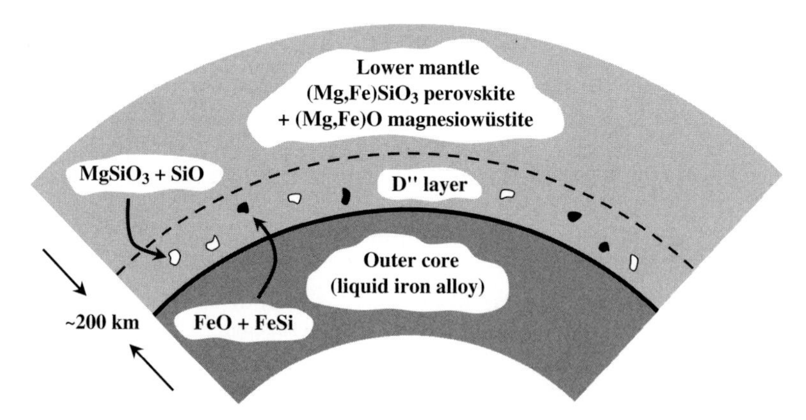

Fig. 7.23. The D″ layer, a layer of heterogeneous composition (symbolized schematically by the *blobs*) at the base of the upper mantle. The compositional heterogeneities are caused by chemical reactions between the core and mantle material

radial) variations, and seismic waves are strongly scattered. The scattering indicates heterogeneities ("blobs" of different compositions) about 10–70 km in size throughout the region.

In laser-heated diamond-anvil cells in the laboratory, molten iron and iron alloys such as FeS are observed to react vigorously with crystalline oxides and silicates such as $(Mg,Fe)SiO_3$ perovskite at temperatures and pressures representative of the core–mantle boundary (4500 ± 1200 K and 1.3 Mbar (130 GPa)).

The overall reaction is

$$Fe + (Mg, Fe)SiO_3 \rightarrow (Mg, Fe)O + FeSi + MgSiO_3 + SiO_2$$

molten iron + perovskite → wüstite + FeSi + (iron-free) perovskite + stishovite

These reactions appear to create a transition region of mixed composition (the D″ layer) separating the molten iron core from the predominantly perovskite lower mantle. Variations in the reactions from place to place may produce the heterogeneities observed.

The D″ layer would allow oxygen, silicon, and magnesium to enter the core and iron to enter the lower mantle. The "rapid" convection currents (~10 km/y) in the outer core then mix the oxygen, silicon, and magnesium through the outer core, implying that the outer core is evolving over time, becoming increasingly contaminated by the mantle.

The reactions occur only at high pressures, >0.3 Mbar (approximately, depending on the material), and so would not be important in the Earth's upper mantle or the mantles of smaller bodies such as Mars and the Moon.

References

Anderson, D. L. 1989. *Theory of the Earth* (Oxford: Blackwell).

Bassett, W. A. 1979. "The Diamond Cell and the Nature of the Earth's Mantle," *Annual Review of Earth and Planetary Sciences*, **7**, 357–384.

Dziewonski, A. M. and Anderson, D. L. 1981. "Preliminary Reference Earth Model," *Physics of the Earth and Planetary Interiors*, **25**, 297–356.

Liu, L.G. 1979. "Phase Transformations and the Construction of the Deep Mantle," in *The Earth: Its Origin, Structure and Evolution*, ed. M. W. McElhinny (London: Academic Press) pp. 117–202.

Plummer, C. C. and McGeary, D. 1988. *Physical Geology*, 4th Ed. (Dubuque, IA: William C. Brown).

Ringwood, A. E. and Major, A. 1971. "Synthesis of majorite and other high pressure garnets and perovskites," *Earth Planet. Sci. Letters*, **12**, 411–418.

8. The Moon's Surface, Structure, and Evolution

Galileo provided the first telescopic description of the Moon. Many subsequent atlases have been created, and lunar orbiters have provided detailed images of nearly all of the Moon's surface.

We list the bulk properties of the Moon in Milone & Wilson (2008), Chapter 13 in the context of the other moons of the solar system, but it is relevant to mention some of its unique characteristics here. The Moon is the largest satellite in the inner solar system, with a radius of 1737 km = $0.2724 R_\oplus$ and mass $0.0123 M_\oplus$. It is the only extraterrestrial body on which people have actually landed and explored the surface. It has also a unique dynamical history, which we will discuss later in this chapter. First, however, we will discuss its composition and appearance.

8.1 Surface Composition

Sources of information about the composition of the Moon include:

1. Ground-based and spacecraft detectors
2. Lunar lander experiments and retrievals

From (1), we gain details from several types of spectrographs and detectors which were targeted at specific types of phenomena: x-ray fluorescence and gamma-ray spectroscopy. Ground-based instruments can reveal the presence of minerals (see Chapter 7), primarily through their infrared signatures.

8.1.1 Lunar Orbiting Spacecraft Detections

8.1.1.1 X-Ray Fluorescence The lunar surface materials absorb solar x-rays. Aluminum (Al), silicon (Si), and magnesium (Mg) in the surface layer then fluoresce, a process that involves the cascading of electrons from higher energy levels, to produce longer wavelength radiation (Figure 8.1). Al, Si, and Mg abundances in the lunar surface materials have been measured to a resolution of about 20 km.

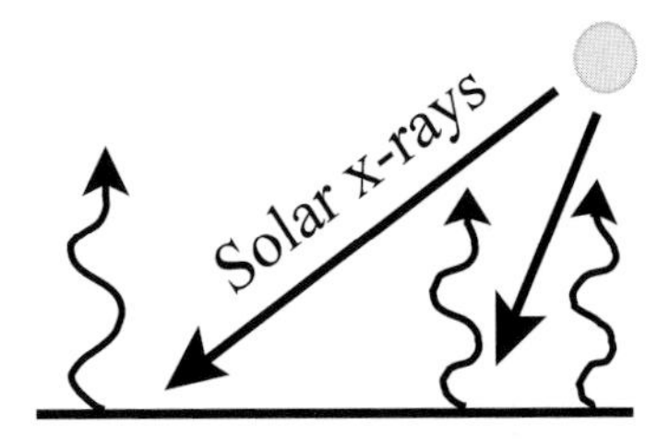

Fig. 8.1. Solar x-rays and fluorescence

Observed ratios of these elements allow a determination of certain major mineral types in the lunar surface layer. Two important examples, their chemical makeup, and main characteristics are:

1. Anorthositic plagioclase feldspar:

   $CaAl_2Si_2O_8$—high Al/Si ratio, low Mg/Si ratio
2. Lunar basalt: pyroxene ($MgSiO_3$) + feldspar—low Al/Si ratio, high Mg/Si ratio

8.1.1.2 Gamma-Ray Spectrometry Gamma rays (frequently written γ-rays) are high-energy photons emitted by the lunar surface. There are two components, arising from different sources:

1. γ-rays emitted by the radioactive decay of uranium (U), thorium (Th), and potassium (K) and their daughter products in the top few centimeters of the lunar regolith
2. Emission resulting from irradiation of the regolith by galactic cosmic ray particles

Detectors in the spectral region 0.54–2.7 MeV reveal mainly the component due to radioactive decay.

8.1.2 Landers (1969–1973)

From the Apollo 11–17 (excluding 13) missions, 382 kg of samples were returned. Heat flow was measured and seismic stations were set up. Three Soviet unmanned Luna landers were also deployed and samples were returned. The Apollo command modules contained detectors that were able to survey the Moon. Figure 8.2 summarizes the surveys' highlights.

The physical variations on the Moon can be viewed in Figures 8.3a–c and are described in the next section.

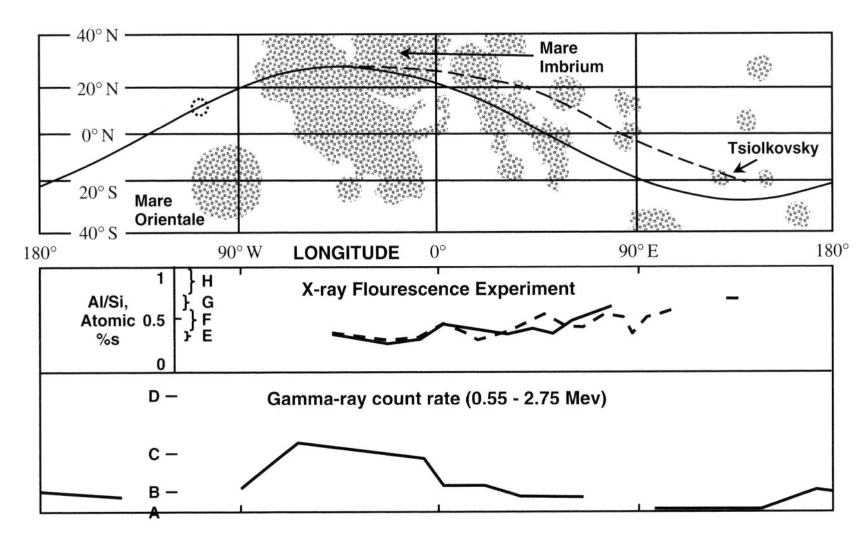

Fig. 8.2. Large-scale compositional variation on the Moon as determined by Apollo 15 orbital experiments. The solid and dashed lines in the lower two plots (depicting the Al/Si ratios from the x-ray fluorescence experiment and the γ-ray counts, respectively) refer to data taken along the solid and dashed trajectories in the map. The γ-ray counts are incompletely reduced here, but include data from earlier Apollo mission. Symbols: A: inert material; B: Apollo 11 Soil; C: Apollo 12 soil 12070; D: Apollo 14 soil; E: mare basalts; F: lunar norites; G: anorthositic gabbros: H: gabbroic anorthosites. The dotted circle on the map shows the approximate spatial resolution of the instruments. From Wood (1972) for the Lunar Sample Analysis Planning Team and adapted with permission from the AAAs

8.2 Lunar Surface Characteristics

The main characteristics of the lunar surface, apparent in Figures 8.3a–c, are:

- Highlands (or *terrae*; light, rough rock), making up more than 80% of the surface.
- Lowlands (mostly dark, flat, roughly circular impact basins) composing ∼16% of the surface

The lowlands include the lunar *impact basins*, which on the near side are lava-filled. Some impact basins exist on the far side, but there is relatively very little lava in them. The lava plains (also called *maria*) cover 35% of the near side.

Next we summarize the main types of surface materials.

Fig. 8.3a. Apollo 11's crew caught this view of the Moon in which the familiar eastern edge is down and near the center of this image, whereas the north limb is to the upper right (compare to the near- and far-side facings in Figures 8.3b and c, respectively). The circular basin to the right of center is Mare Crisium, and that to the left of center is Mare Fecunditatis. Mare Tranquillitatis is above it, and near the upper limb is Mare Serenitatis. The lunar Pyrenees separate M. Fecunditatis from Mare Nectaris on the left, above the bright-rayed craters. NASA photo AS11-44-6665

8.2.1 Regolith

The lunar surface has been pulverized by impacts. The evidence is to be found in the powdery layer of pulverized rock known as *regolith.* Its depth over the surface is:

- 2–8 m on the maria
- Up to 15 m, or more, in the highlands

8.2.2 Breccias

Generally, *breccia* is rock formed when rock fragments are welded together by metamorphism. Most lunar rocks are, in fact, breccias:

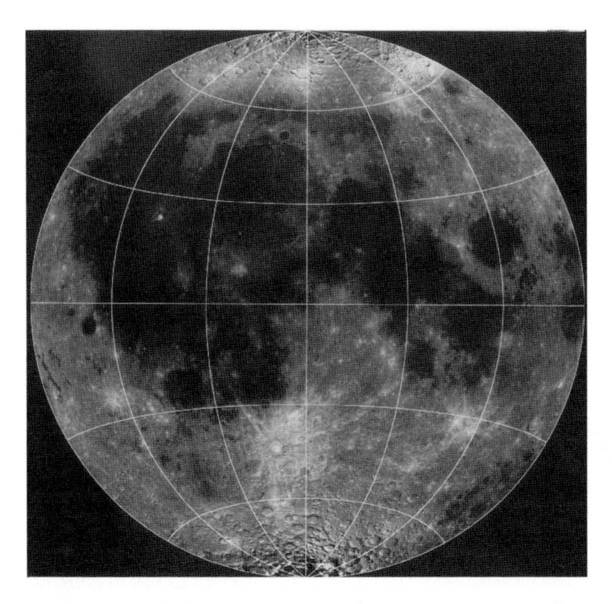

Fig. 8.3b. The US Geological Survey albedo (relative brightness) map of the near side of the Moon produced with data from the Clementine lunar mission at a wavelength of 0.75 μm. North is up and East is to the right. The prominent bright-rayed crater nearly on the central longitude is Tycho, and the prominent dark-floored crater much farther north is Plato. Mare Crisium is to the right. Note the darkness of the "seas" compared to the highlands. Reproduced, courtesy, US Naval Research Laboratory and US Geological Survey

- Regolith welded together by heating during impacts
- Rock fractured by impacts and welded together by impacts

Breccias often contain fragments of older breccias; some lunar rocks show as many as four generations of breccias. Some meteorites, too, are brecciated (Milone & Wilson 2008, Chapter 15.3.1).

8.2.3 Impact Melts

These are fine-grained crystalline rocks created during impacts, which generate enough heat to melt some of the rock at the impact site. All non-brecciated rocks found on the lunar surface are solidified impact melts. They are *not* ancient rocks which have escaped impact brecciation.

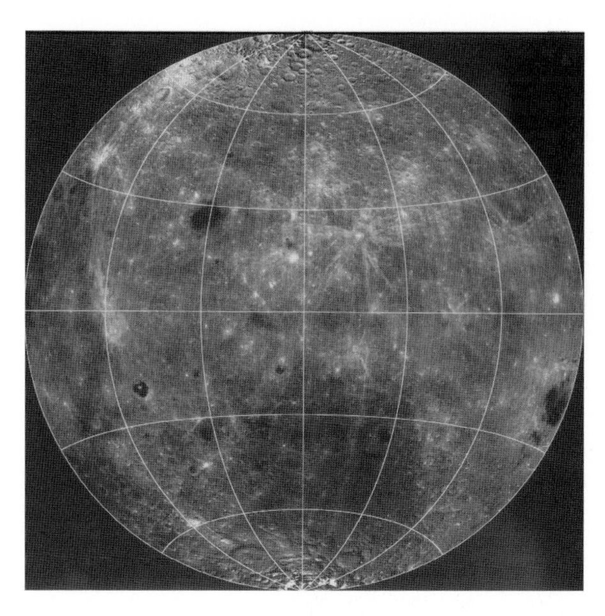

Fig. 8.3c. The US Geological Survey (USGS) albedo (relative brightness) map of the far side of the Moon produced with data from the Clementine lunar mission at a wavelength of 0.75 μm. North is up and East is to the right. Note the relative paucity of maria. The dark crater with brilliant white peak, located at ~50°W longitude and −20°S, is Tsiolkovsky. Reproduced, courtesy, US Naval Research Laboratory and US Geological Survey

8.2.4 Terrae (Highlands)

The surface has been totally pulverized by impacts. As a consequence, there are no bedrock exposures.

8.2.4.1 Ages of Highland Rocks From radioisotope measurements (see Milone & Wilson, 2008, Chapter 15.5.1), the ages of lunar rocks are found to be mostly 4.0–3.8 Gy, due to "resetting" of many age clocks by impacts; but some rocks have ages approaching 4.5 Gy, and one sample of dunite (almost pure olivine) has an age of 4.6 Gy.

The interval 4.0–3.8 Gy is very brief in geological terms. Is it due to a short-lived high-impact era? Or does it represent simply the tail end of a protracted impact era extending from the accretion of the moon?

8.2.4.2 Dominant Highland Mineral One mineral has a commanding presence on the lunar surface:

Anorthositic plagioclase feldspar, $CaAl_2Si_2O_8$ (see Chapter 7.2.8). This is generally 90–98% anorthite (Ca, Al-rich). Note that the name *plagioclase* covers a complete series: $NaAlSi_3O_8$–$CaAl_2Si_2O_8$, from *albite*, $NaAlSi_3O_8$, to *anorthite*, $CaAl_2Si_2O_8$; in between are *oligoclase*, *andesine*, *labradorite*, and *bytownite*.

Lunar rocks also contain up to 10 times more titanium than terrestrial rocks. *Ilmenite* ($FeTiO_3$) is a common mineral containing this element. These data are important clues to the origin and evolution of the Moon.

8.2.4.3 Highland Rock Types

1. The most abundant rock type is highland basalt. Its makeup is, typically:

   70% anorthositic plagioclase feldspar
   20% *orthopyroxene* ($MgSiO_3$–$FeSiO_3$ series)
   9% *olivine* [$(Mg, Fe)_2SiO_4$; i.e., a complete series between Mg_2SiO_4 and Fe_2SiO_4]
   1% *ilmenite*, $FeTiO_3$

2. The second most abundant is low-K Fra Mauro basalt, typically composed of:

   53% anorthositic plagioclase feldspar
   30% orthopyroxene ($MgSiO_3$–$FeSiO_3$ series)
   10% olivine
   7% *clinopyroxene* ($CaMgSi_2O_6$–$CaFeSi_2O_6$ series)
   2.5% ilmenite, $FeTiO_3$

   (NB: these add up to 102.5% of the rock, so there may be some inaccuracy in the percentages.)

   Some highland rocks are almost pure anorthite.

   These two types of basalt cannot have come from the same parent magma, so at least two magma types may have coexisted in the early moon.

3. Another rock type found in the highlands is *KREEP* basalt.

   The acronym stands for its enhanced concentrations of: potassium (K), rare earth elements (e.g., samarium, Sm), and phosphorus (P). KREEP basalt is a component of many highland regoliths, soils, and impact melts.
   Lunar basalts occur in surface lava flows, and originate in a molten or partially molten source region below the anorthositic feldspar crust.
   KREEP basalts from different areas have remarkably uniform trace-element abundances and a uniform age of 4.35 Gy. This implies a uniform and short-lived source region.

8.2.5 Maria

The maria are basaltic lava plains. They are very thin:

A few hundred meters thick on average
2–4 km thick in the centers of the big impact basins

8.2.5.1 Mineral Composition of Mare Basalts

35–68% clinopyroxene ($CaMgSi_2O_6$–$CaFeSi_2O_6$ series)
10–40% anorthositic (Ca, Al-rich) plagioclase feldspar:
the plagioclase is 60–98% anorthosite, $CaAl_2Si_2O_8$
0–20% olivine

The albedos of maria basalts are low. They are dark compared to the highlands (about half the albedo of the highlands) because:

They have higher iron content (due to greater pyroxene/olivine content)
Impact melting produces glass spherules as part of the regolith; glass is darker when the iron content is higher

8.2.5.2 Ages of the Maria

The impact events date from the period 4.4–3.8 Gy ago.
The lava flows filling the impact basins date from 3.7 to 3.1 Gy ago.

8.2.5.3 Orange Glass The Apollo 17 astronauts found small glass spherules colored orange due to their high titanium content. They are likely to have originated as lava droplets which were molten when thrown out by an eruption, and solidified in flight. Some were coated with volatiles, e.g., Zn, Pb, S, Cl.

8.2.6 Overall Lunar Composition

Briefly summarized, the abundances of

Fe, Si, Mg are similar to those of Earth's mantle.
Volatiles (materials which are easily evaporated) on the moon as a whole are impoverished compared to the Earth.
Refractory elements (which are not readily evaporated) may be enhanced compared to Earth; titanium is definitely enhanced.

8.2.7 Crater Characteristics

Craters are a principal characteristic of the lunar surface, as Figures 8.3a–c demonstrate. The main points about them are:

- Number: $\sim$300,000 $\gtrsim$ 1 km diameter
- Relief (highest–lowest points): $\sim$2 km

- The presence of many "ghost" craters (those in which only the rim is visible, the craters being filled in by lava and regolith)
- Ray systems
  - Chains of bright, secondary craters and ejecta
  - The most spectacular example, the ray system associated with the crater Tycho, extends over 10,000 km (see Figure 8.3b). According to Baldwin (1963), the radial extent of a ray, R_r, of a large crater is related to the crater radius, R_c, by:

$$R_r = 10.5\, {R_c}^{1.25} \tag{8.1}$$

  where both distances are expressed in km.

- Schröter's law, named for J. H. Schröter (1745–1816), that the volume above = the volume below the surrounding terrain level (i.e., that the volume of material on the rim and beyond is equal to the volume of the excavated crater), may not strictly apply, because the former may exceed the latter significantly (Melosh 1989, p. 90). The thickness of ejecta blanket, δ, in terms of the distance from the crater, r, and the crater radius, R, is suggested by McGetchin et al. (1973) to be

$$\delta = 0.14\ R^{0.74}(r/R)^{-x}, \tag{8.2}$$

  for $r \geq R$, where $2 \leq x \leq 3.7$, as found from small-scale (sandbox) explosions, and roughly 3 for lunar craters.
- Diameter vs. depth relation (from Baldwin, 1963):

$$\log D = 0.103\ (\log d)^2 + 0.803\ \log d + 0.62 \tag{8.3}$$

  where D is the diameter; d is the depth, in km.
- Crater size relation to energy:
  - The velocity of impact may be computed by adding in quadrature the orbital velocities of the Moon and the impactor, and the escape velocity of the Moon (Chapter 5.5.1). The latter is

$$v_\infty = [2GM/R]^{1/2} = 2375\,\text{m/s} \tag{8.4}$$

    where $R = 1.7374 \times 10^6$ m, $M = 7.3483 \times 10^{22}$ kg.

    Whence, assuming identical orbital velocities, we get

$$\frac{1}{2}\,{v_\infty}^2 = 2.82 \times 10^6\ \text{J/kg} = 2.82 \times 10^{10}\ \text{erg/g}$$

– Melosh (1989) suggests, more generally, that $D \propto E^{1/3}$ is adequate as an approximation for smaller craters, and the expression $D \propto E^{1/4}$ for larger lunar craters. The dependence on the density ρ and gravitational acceleration g, of the object being impacted, he summarizes as:

$$D = [fE/(\rho g)]^{1/4} \tag{8.5}$$

where f is a constant fraction. For terrestrial craters, he suggests an expression such as:

$$D = 1.8\rho_{\mathrm{i}}^{0.11}\,\rho_{\mathrm{p}}^{-1/3}\,g_{\mathrm{p}}^{-0.22}\,R_{\mathrm{i}}^{0.13}\,E_{\mathrm{i}}^{0.22} \tag{8.6}$$

where the subscripts i and p refer to the impactor and the planet (or moon, *target* generally), respectively (Melosh 1989, p. 121).

The relation between the energy of impact and the seismic energy radiated (in Joules) from the impact is given by

$$E_{\mathrm{seismic}} = kE, \tag{8.7}$$

where the *seismic efficiency* is in the range $10^{-5} \leq k \leq 10^{-3}$, and where the seismic energy is related to the Gutenberg–Richter earthquake magnitude, M, through

$$\log E_{\mathrm{seismic}} = 4.87 + 1.5\,M \tag{8.8}$$

or

$$M = 0.67 \log E_{\mathrm{seismic}} - 4.87$$

With $k = 10^{-4}$, a 30-m diameter impactor striking the Moon at 20 km/s would produce an $M = 5.6$ quake; one hitting the Earth would produce a 1 km diameter crater.

Melosh (1989, esp. Ch. 5) discusses in detail the formation of craters as a function of time, and is an essential reading for this area of study.

8.3 Seismic Results

P-waves propagate throughout the Moon, with speed changes (see Figures 8.4 and 8.5). The S waves dissipate by a depth of 700–800 km (Figure 8.6).

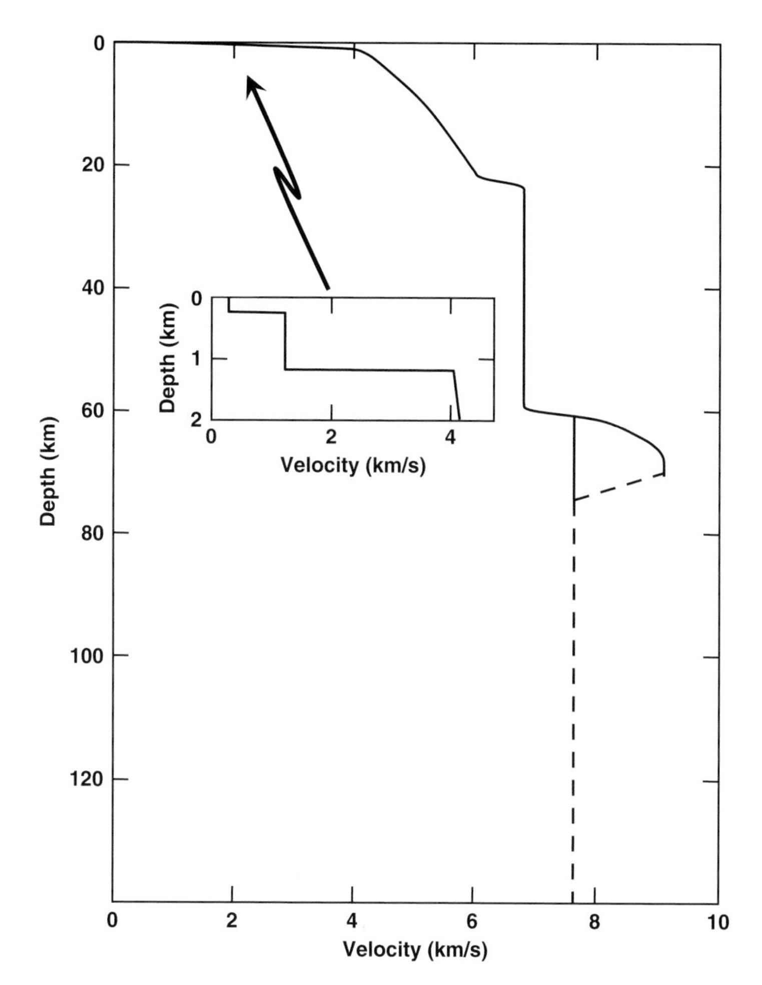

Fig. 8.4. P-wave speed as a function of depth into the Moon. Based on Toksöz et al. (1973, Fig. 6, p. 2536) and Taylor (1975, Fig 6.7, p.290)

8.3.1 Upper 25 km

In the Mare Serenitatis region (from Apollo 17 experiments):

To ~250 m depth, pressure-wave velocities undergo a change:

$$v_{\mathrm{p}} \approx 200\text{–}300\,\mathrm{m/s}\ (0.2\text{–}0.3\,\mathrm{km/s}).$$

At ~250 m depth, there is a sudden change to 1 km/s.
A similar change is seen in terrestrial lava flows, with no change in rock type.
At 1.2 km depth, v_{p} changes abruptly to 4 km/s.

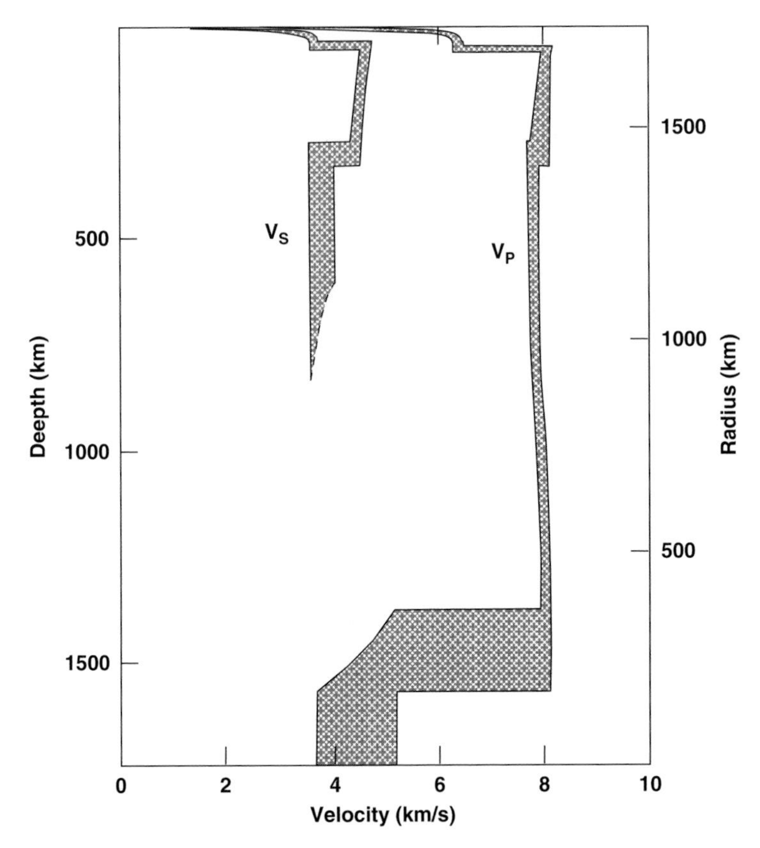

Fig. 8.5. S-wave and P-wave speeds as a function of depth into the Moon. Note that the S-waves are highly attenuated below 700 km depth. Adapted from Nakamura, et al. (1974, Fig. 3, p. 139). Copyright 1974 American Geophysical Union

This is interpreted as the base of the mare basalts and marks a transition to the underlying highland basalt.
This is followed by a slow increase to 6 km/s at 25 km depth.

8.3.2 25-km Discontinuity

At this depth, there is a sharp increase from 6 to 6.8 km/s over ~ 1 km in depth, followed by a slow increase to 7 km/s between 25 and 60 km depth.

Apparently the composition is uniform with material above 25 km; but at this level a change in physical characteristics probably occurs. Perhaps impact fracturing did not reach below 25 km or fractures may be self-annealing below 25 km. If the latter, it is not due to pressure-closing of cracks, because the

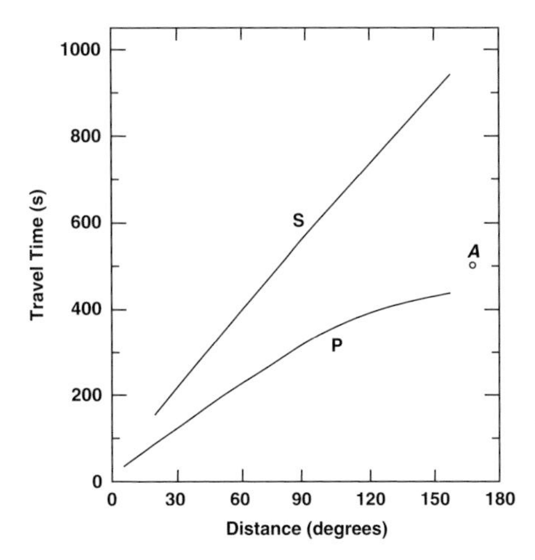

Fig. 8.6. Travel time curves for S-waves and P-waves from seismic sources on the surface of the Moon, measured in degrees (180° is antipodal from the source). The point labeled A is a single observation that suggests the existence of a lunar core. Based on Cook (1980, Fig. 5.8a, p. 150) and Nakamura et al. (1974, Fig. 1, p. 138). Copyright 1974 American Geophysical Union

pressure at 25 km in the Moon is only $\sim$1.2 kbar, and this is not enough to close the cracks; one would also expect pressure closing to occur gradually with increasing depth (i.e., with increasing pressure).

8.3.3 25–60 km Depth

Over this range of depth, the observed v_{p} suggests lunar highland anorthositic gabbros. Gabbro is the plutonic equivalent of basalt (solidified at depth). The anorthositic crust is thicker on the far side, by $\sim$12 km, in average. This may contribute to an offset between the geometric center and the center of mass of the Moon of about 1.8 km.

8.3.4 60-km Discontinuity

Here, v_{p} increases from 7 to $\sim$8 km/s. It marks a transition from anorthositic crust to olivine/pyroxene mantle below 60 km ($\rho \approx 3400\,\mathrm{kg/m^3}$). In this context, the term anorthositic means that the material has a high content of anorthositic plagioclase feldspar. This implies a basaltic composition above 60 km. See Figure 8.4.

There is a possible high-velocity zone (HVZ) immediately below the crust, but if it truly exists, it must be localized.

8.3.5 Lithosphere

8.3.5.1 Upper Mantle The upper mantle is 250 km thick (60 to ~300 km depth). Its characteristics are:

$v_p \approx 8.1$ km/s; $v_s \approx 4.2$ km/s;
75–80% olivine, 20–25% pyroxene, $Mg_{0.8-0.9}$ $Fe_{0.2-0.1}$, suggesting a mantle composition similar to Earth.

8.3.5.2 Middle Mantle The middle mantle is about 700 km thick (300–1000 km depth). Its characteristics are:

$v_p \approx 8.1$ km/s but $v_s \approx 3.8$ km/s, a slower S-wave speed than in the upper mantle.

8.3.6 Asthenosphere

8.3.6.1 Lower Mantle The lower mantle may extend from a depth of 1000 km to the center of the Moon. But this is highly uncertain. S-waves are strongly attenuated (die out before they can pass through it). This implies that it may be at least partially molten. Figure 8.5 illustrates this model.

8.3.6.2 Possible Core The only direct evidence known to us for a core is a single seismic observation at 168° (almost antipodal from the seismic source). This implies that the P-waves would have traveled through the core *if* there is a core at all. The waves arrived late compared to what one would expect by extrapolating the travel time curve observed at smaller distances from the source. Therefore the P-waves appeared to slow down in the region interpreted as a "core" ($v_p \sim 3.7$–5.1 km/s). The observation implies a core ~170–360 km radius of undetermined composition. See Taylor (1999) for further discussion of lunar structure.

8.4 Geochemical Evolution

8.4.1 Formation of the Asthenosphere: Three Scenarios

The asthenosphere is the partially molten inner part of the Moon, below a depth of ~1000 km ($R < \sim 700$ km).

One can imagine at least three scenarios for the formation of the asthenosphere.

8.4.1.1 Rapid Accretion of the Moon If the Moon accreted in only $\sim$ 100–1000 years, then it would have become totally impact-melted. Under this scenario, iron would sink to the center, creating an Fe–FeS core with these properties:

- $R_{\text{CORE}} \lesssim 700\,\text{km}$
- Molten at $T \gtrsim 1000\text{–}1100^\circ\,\text{C}$

A dynamo in this early core could then perhaps have created the ancient magnetic field of the Moon.

This may account for the depletion of the siderophile (iron-loving) and chalcophile (copper-loving) elements in the crust, *if* the Moon contained this Fe and S in the first place.

But they may not have been present in the first place. If not, then the geochemical need for a core disappears.

8.4.1.2 Longer Accretion Time for the Moon If only the outer 90% of the Moon's volume became impact-melted, then the asthenosphere may be undifferentiated, primitive material. In this case, the partially molten state of the asthenosphere may be due to trapped U, Th, and K. The seismic data are satisfied by 0.5–1% partial melting.

8.4.1.3 Rapid Accretion of the Moon with Low Fe, S Abundance Then the asthenosphere may be the remnant of the early molten interior, and is still cooling.

Thus, low Fe–FeS content $\Rightarrow$ no actual core.

8.4.2 Depth of Magma Ocean

With a deep ocean, we might expect all volatiles to evaporate. But Apollo 17 found volatile-coated orange glass beads. This discovery suggests that pockets of volatile-rich material survived, which may imply that the magma ocean was no more than a few hundred km deep.

8.4.3 Initial Fractionation (First 100 million yr, 4.6–4.5 Gy)

Suppose we assume a model as in Section 8.4.1.2, for example, starting with a Moon 90% molten, with the inner 700 km of the radius not melted. Then the scenario could go like this.

8.4.3.1 First Stage

Olivine (Mg_2SiO_4-rich) ⇒ 2 Mg ions precipitate for every 1 Si
Olivine is dense so it sinks
The remaining melt becomes progressively depleted in Mg compared to Si

8.4.3.2 Initial Crust Rapid cooling at the surface ⇒ olivine/pyroxene crust.

8.4.3.3 Second Stage The second stage is characterized by a decreasing Mg/Si ratio: it becomes increasingly difficult to precipitate a mineral containing 2 Mg per Si (olivine), so there is increasing precipitation of orthopyroxene ($MgSiO_3$, Mg:Si=1:1).

Thus olivine and orthopyroxene (Ol–Opx) precipitate together. Olivine and orthopyroxene include Mg, Ne, Fe, Co, and Cr^{2+} ions, but exclude Ca, Al, and others. This means there was an increasing Ca, Al concentration in the residual melt. The overall picture is shown in Figure 8.7.

The model depicted in Figure 8.7 accounts for the overall structure of the Moon, but not for the asymmetries, such as the thinness of crust on the near side compared to the far side, or the offset of center of figure from center of mass. The thickness of the crust on the far side is consistent, though, with the lack of maria.

Positive Bouguer gravitational anomalies (*mascons*) have been detected under the basalt lava plains on the near side, amounting to +220 mgals in Mare Imbrium, and +200 mgals in Mare Orientale, a ringed plain. Most younger craters under 200 km diameter show negative anomalies because of the material excavated. That of the Sinus Iridium, the "bay" near the South Pole, is −90 mgal.

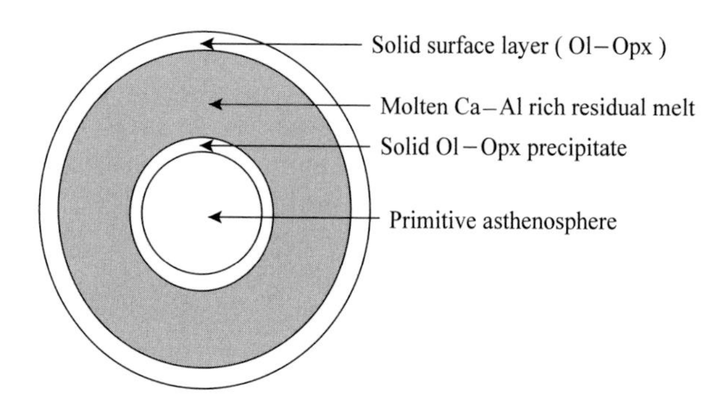

Fig. 8.7. Initial fractionation scenario for the Moon

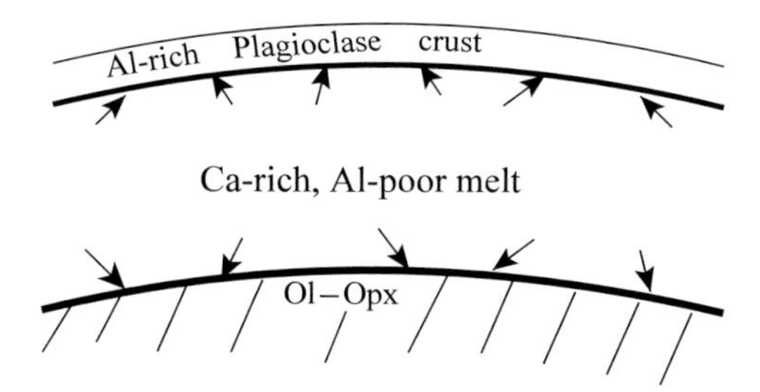

Fig. 8.8. Precipitation and compositional variation in the Moon

8.4.3.4 Third Stage When the aluminum concentration is $\gtrsim 12\%$ Al_2O_3 (*corundum*), anorthositic plagioclase feldspar precipitates. It is less dense than the residual melt and therefore had to rise, forming the primary crust (now seen in the lunar highlands).

Impact mixing totally mixed the initial Ol–Opx crust with this later feldspar-rich crust. The single collected lunar dunite rock (4.6 Gy age) may be a remnant from the initial crust.

The olivine and pyroxene continue to precipitate and sink, but with increasing concentration of clinopyroxene ($CaMgSi_2O_6$) because of the high Ca content of the residual melt. The process is illustrated in Figure 8.8.

This continues until the base of the melt is about 300 km deep (Moon 60% solidified), with a 60-km thick crust.

8.4.3.5 Final Solidification The last of the residual melt (60–300 km depth) cools and solidifies. It is Ca-rich, Al-poor (high Ca/Al ratio) because of the upward precipitation of plagioclase feldspar into the crust. The end result is that there is considerable zoning as the final melt solidifies. Thus, the composition varies with depth.

U, Th, and K have large ion radii. The implications of this for lunar evolution are:

- That they are difficult to incorporate into minerals, so they became concentrated in the last region to solidify (60–300 km depth).
- Their radiogenic heating later produced partial melting.
- Subsequently, lava from this basalt source region flowed onto the surface.

The partial melting zone migrated downward (regions with lower U, Th, K concentrations took longer to melt); following this, a succession of basalts erupted with different compositions over time.

8.5 Dynamical History of the Moon

The orbit of the Moon is inclined slightly (5.09°) to the ecliptic, not the equator, suggestive more of a binary companion to, rather than a satellite of, the Earth. Its orbit has a perceptible eccentricity (0.0549), sufficiently large that the angular size of the Moon can be seen to vary by ∼11% from perigee to apogee. Thus, although the mean center-to-center distance from Earth is 384,400 km, the Moon's distance varies between 363,000 and 406,000 km. Moreover, because of perturbations, the longitude of the nodes regresses with a period of ∼18.6 years, and the perigee advances with a period of ∼9 years. The nodal regression causes a large variation in the extreme declinations of the Moon and thus of the extreme northern and southern azimuths of moonrise, especially at high latitude observing sites. See Kelley and Milone (2005), for a full discussion of this effect and its possible observation in Neolithic times and in antiquity.

The semimajor axis also varies. It appears to be increasing secularly, at a rate of ∼4 cm/y (see Stephenson 1997 for a full discussion), most likely due to tidal friction on Earth and the acceleration of the Moon as a consequence of a tidal bulge on Earth preceding the sublunar longitude. The tidal forces are proportional to the inverse cube of the distance; consequently, the prediction is that the recession of the Moon will continue until the solar tides dominate.

Extrapolating into the past, the Moon must have been much nearer to the Earth, when the Earth had a shorter period of rotation. The evidence of stromatolite growth patterns has suggested a much shorter rotation period for Earth and a shorter month as recently as the Devonian.

Various scenarios for the Moon's origins have been put forward since Darwin's (1880) hypothesis that the Moon spun off from the Earth when the two objects were formed with too high angular momentum. His

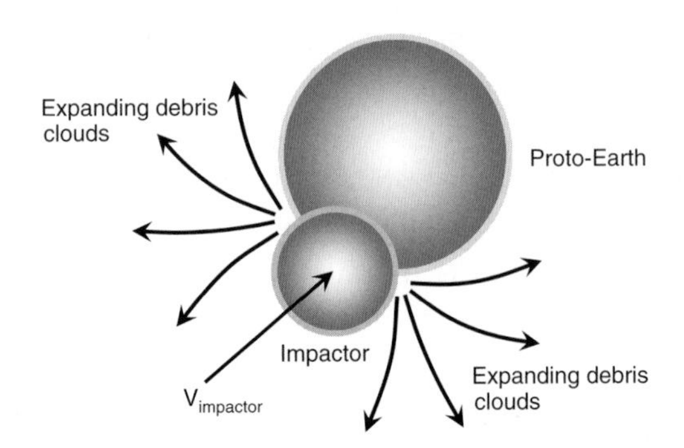

Fig. 8.9. A major impact by a Mars-sized body on a proto-Earth is currently the favored theory for the origin of the Moon. Adapted loosely from Stevenson (1987, Fig. 1A, p. 274) and appearing here with permission of Annual Reviews

hypothesis, modified by Ringwood (1966), accounted to some extent for the similarities of composition between the Earth and the Moon, but failed to account for the difference in chemical composition between them. Moreover, it is not clear how the Earth–Moon system could form with a break-up amount of angular velocity. Two other hypotheses, namely that the Earth captured the Moon (e.g., Öpik 1972) and that the Earth and Moon somehow were formed near each other as a true binary planet (Ruskol 1960) similarly have failed to satisfy fully both dynamical and compositional property differences and similarities. A nice discussion of the difficulties with these three

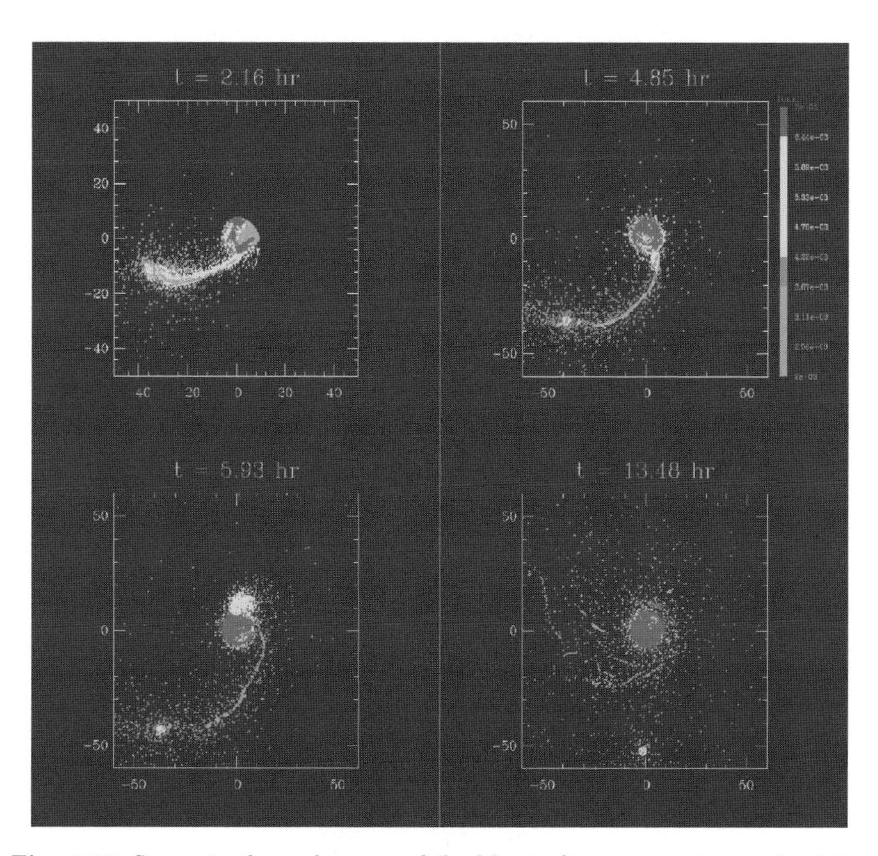

Fig. 8.10. Stages in the coalescence of the Moon after a major impact involving a proto-Earth in the early solar system. Much of the material falls back to Earth, a small amount escapes the system, and some goes into orbit around the Earth to form a disk that later coalesces into the Moon. The clump at 6 o'clock in the lower right panel is on an eccentric orbit that later passes within the Earth's Roche limit (see Milone & Wilson 2008 Chapter 13.2), so that this clump, itself, is not a candidate for the Moon. Note that simulated temperatures vary from ~2000 K for many fragments to ~7000 K at the Earth. From a simulation by Robin Canup, and kindly provided by her (see Canup, 2004 for a full discussion) and appearing here with the permission of Elsevier. (See also Plate 5)

hypotheses and discussion of the alternative, major impact theory described below can be found in Stevenson (1987).

The strongest and most widely accepted formation hypothesis currently involves a massive impact by a Mars-sized body on an early Earth (Cameron 1985, Benz et al. 1986; Canup (2004)). Models of such an event are illustrated in Figure 8.9 and with detail in Figure 8.10, where Robin Canup has simulated the collision over time intervals following the catastrophic collision. Such models appear to be able to account for both similarities and differences in composition by having the dense core of the Mars-like body sink within that of the proto-earth, and the mantle only partially mixing with Earth while most of the rest forms the Moon. This scenario helps to explain also the relatively low density of the Moon (see Section 5.2), and its small or absent iron core. It also explains why the Moon seems to be deficient in volatile as opposed to refractory material (i.e., it has a paucity of material that is slowest to condense as the temperature of the environment decreases, but an abundance of material that condenses earlier). It should be noted, however, that Stevenson (1987, pp. 277–278) suggests that the degree of deficiency of the volatiles in the Moon is uncertain. Dynamically, the models help to explain the recession of the Moon and why the chemical composition, although different from the Earth's, is more like that of Earth than of anything else in the solar system.

This discussion will be taken up again in Chapter 15, when we discuss the evidence of the meteorites and the origin of the solar system.

References

American Association for the Advancement of Science 1970. "The Moon Issue," *Science*, **167**, No. 3918 (January 30, 1970).

Baldwin, R. B. 1963. *The Measure of the Moon* (Chicago, IL: University of Chicago Press).

Cameron, A. G. 1985. "Formation of the Prelunar Accretion Disk," *Icarus*, **62**, 319–327.

Canup, R. 2004."Simulations of a late lunar-forming impact," *Icarus*, **168**, 433–456.

Cook, A. H. 1980. *The Interiors of the Planets* (Cambridge: Cambridge University Press).

Darwin, G. H. 1880. "On the Secular Change in Elements of the Orbit of a Satellite Revolving around a Tidally Distorted Planet," *Philosophical Transactions of the Royal Society of London*, **171**, 713–891.

Kelley, D. H. and Milone, E. F. 2005. *Exploring Ancient Skies: An Encyclopedic Survey of Archaeoastronomy* (New York: Springer-Verlag).

McGetchin, T. R., Settle, M., and Head, J. W. 1973. "Radial Thickness Variation in Impact Crater Ejecta: Implications for Lunar Basin Deposits," *Earth and Planetary Sciences Letters*, **20**, 226–236.

Melosh, H. J. 1989. *Impact Cratering: A Geologic Process* (New York: Oxford University Press; Oxford: Clarendon Press)
Milone, E. F. and Wilson, W. J. F. 2008. *Solar System astrophysics: Planetary Atmospheres and the Outer Solar System.* (New York: Springer).
Nakumara, Y., Latham, G., Lammlein, D., Ewing, M., Duennebier, F., and Dorman, J. 1974. "Deep Lunar Interior inferred from Recent Seismic Data," *Geophysical Research Letters*, **1**, 137–140.
Öpik, E. J. 1972. "Comments of Lunar Origin," *Irish Astronomical Journal*, **10**, 190–238.
Ringwood, A. E. 1966. "Chemical Evolution of the Terrestrial Planets." *Geochimica Cosmochimica Acta*, **30**, 41–104.
Ruskol, E. L. 1960. "Origin of the Moon. I," *Soviet Astronomy, AJ* **4**, 657–668.
Stephenson, F. R. 1997. *Historical Eclipses and Earth's Rotation* (Cambridge: University Press).
Stevenson, D. J. 1987. "Origin of the Moon—The Collision Hypothesis," *Annual Review of Earth and Planetary Sciences*, **15**, 271–315.
Taylor, S. R. 1975. *Lunar Science–A Post-Apollo View* (Oxford: Pergamon press).
Taylor, S. R. 1999. "The Moon" in *Encyclopedia of the Solar System*, ed. P. R. Weissman, L.-A. McFadden, and T. V. Johnson, pp. 247–275.
Wood, J. A., (Lunar Sample Analysis Planning Team). 1972. "Third Lunar Science Conference: Primal Igneous Activity in the Outer Layers of the Moon Generated a Feldspathic Crust 40 kilometers Thick," *Science*, **176**, No. 4038, 975–981.

Challenges

[8.1] Calculate the lunar crater diameter expected from the impact of an object with a 100 m diameter and density of 3400 kg/m^3. Suppose that the object is overtaken by the Moon with a net orbital speed difference of 5 km/s.

[8.2] The crater frequency is higher over the highlands than over the lowlands of the Moon. Why? Discuss the situation in the light of the time-line of selenological history.

[8.3] The moment of inertia of the Moon is 0.391; what does this imply about the structure of the Moon?

[8.4] The angular momentum of the Earth–Moon system is 3.41×10^{34} kg m^2/s. Assuming no loss of angular momentum to the Earth-Moon system since then, find the rotation speed of the two bodies when they were in contact, under the *fission theory* for the Moon's origin. What can you conclude assuming the *impact theory* for the Moon's origin?

[8.5] If the acceleration of gravity is matched by the centrifugal acceleration in a critically rotating contact Earth–Moon binary, compute the break-up speed. Is the speed computed in [8.4] sufficient for fission to have occurred?

9. Surface Science of the Terrestrial Planets

The orbital, physical, and photometric properties of the terrestrial planets are summarized in Table 9.1. In this chapter, we target the surfaces as well as the interiors of Mercury, Venus, and Mars, and compare their properties to those of the Earth–Moon system, which we have already examined. In Milone & Wilson (2008, Chapter 10), we examine the nature of atmospheres and ionospheres with tools of physics and chemistry. We consider the planets in their heliocentric order, starting with Mercury.

9.1 Mercury

9.1.1 Visibility

Mercury is the smallest of the terrestrial planets[1] with a radius of 2439 km, or 0.38 $\Re_{\oplus}$. The planet was known in antiquity as the messenger of the gods, because of its apparent rapid shuttle motion back and forth around the Sun, overtaking planet after planet during the course of its elaborate motions. In ancient Greece it was known as Hermes when seen in the evening sky, and Apollo when seen before sunrise. In the ancient Egyptian culture the two apparitions were sometimes given the names Horus and Seth, and among the Hindus, Raulineya and Buddha.

Its angular diameter varies between about 5 and 11 arc-secs as viewed from Earth. At maximum elongation, a typical brightness is ~0.5 magnitude and a diameter of ~8 arc-secs.

Mercury is difficult to see because of its proximity to the Sun. Its maximum elongation is ~28° (see Figure 9.1). Abell (1975, p. 302) commented that despite the fact that it is the seventh brightest object (after the Sun, Moon, Venus, Mars, Jupiter, and Sirius) in our sky, "... most people—including even Copernicus, it is said—have never seen Mercury."

[1] It is also the smallest planet, overall, if Pluto, which is smaller than Earth's moon, and no longer thought to be the largest icy body at the outer fringes of the solar system, is excluded (see Milone & Wilson 2008, Chapters 13–16).

Table 9.1. Bulk properties of the terrestrial planets

Properties	Mercury	Venus	Earth	Mars
Orbital data				
a (au)	0.3871	0.7233	1.0000	1.5237
e	0.2056	0.0068	0.0167	0.0934
i (to ecliptic)	7°.0047	3°.3947	0°.0007	1°.8494
Sidereal period	0y.2408	0y.6152	1y.0000	1y.8809
Physical data:				
Mass ($M_\oplus$)	0.0553	0.8150	1=5.974 × 10^{24} kg	0.1074
R_{eq} ($R_\oplus$)	0.3825	0.9488	1=6378.137 km	0.5325
$P_{rotation}$	58.6462	−243.0185	0.99726963	1.02595676
(Sidereal, in mean solar days)				
Oblateness[a]	0	0	0.0035364	0.006772[b]
J_2	...	2.70 × 10^{-5}	1.082 × 10^{-3}	1.964 × 10^{-3}
J_3	...	...	−2.54 × 10^{-6}	3.6 × 10^{-5}
J_4	...	...	−1.61 × 10^{-6}	...
$<\rho>$ (kg/m^3)	5430	5240	5515	3940
$<g_{eq}>$ (m/s^2)	3.703	8.871	9.798	3.711
$V\infty$ (km/s^1)	4.25	10.36	11.18	5.02
Photometric data:				
V (1,0)	−0.42	−4.40	−3.86	−1.52 [−2.01][c]
(B - V)	+0.93	+0.82	...	+1.36
(U - B)	+0.41	+0.50	...	+0.58
Radiative energy data:				
Albedo (geom., V)	0.106	0.65	0.367	0.12
Incident solar flux (W/m^2)	9076	2600	1360	586
T_{eq} (predicted effective temperature)	517 K[d]	299 K[d]	248 K[e]	216K[e]
T_{eff} (observed)	700 K[f]	740 K	288 K	210 K

[a] $\epsilon = (R_{eq} - R_{pol})/R_{eq}$.
[b] N pole result; S pole result: 0.005000.
[c] [V_0, the photometric V magnitude at mean opposition]. V(1,0) is the V magnitude that the planet would have if it were viewable at 1 au from the Sun and 1 au from the Earth, at a phase angle of 0 (see chapter 5.5.2).
[d] "slow rotator" case; based on visual albedo.
[e] "fast rotator" case; based on visual albedo [(for $A_\oplus$ (bol) = 0.307, $T_\oplus$ = 254 K)]
[f] subsolar temperature at perihelion

Its computed magnitude at superior conjunction (when it is not readily visible) is −2.2, which potentially makes it even brighter than Sirius. Its proximity to the Sun means that it must be observed in daylight, or twilight, and, if it is to be observed when the Sun is below the horizon, it is seen through high air mass. Not surprisingly, the best ground-based

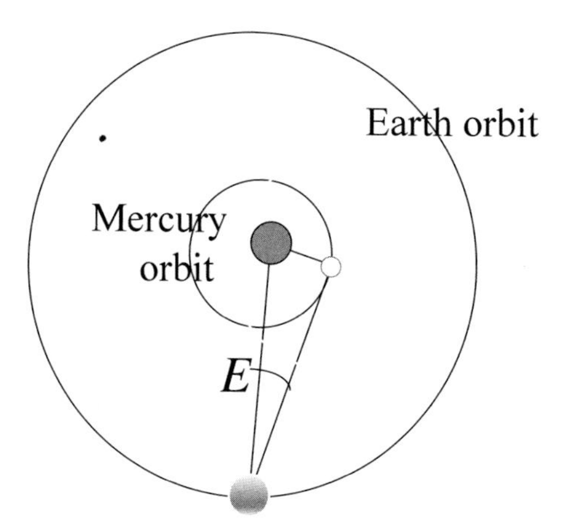

Fig. 9.1. Mercury, at a "maximum western" *elongation* (E). As the earth rotates CCW, Mercury in this configuration rises before the Sun, hence west of the Sun in our sky. Note that this particular maximum elongation is not optimal for obser-vating Mercury

views have not been especially good. Consequently, although Mercury has been photographed on occasion, the best views we have of Mercury are from spacecraft.

9.1.2 Mercury's Orbit

Mercury has a semi-major axis of 0.38710 au and the inclination of its orbit to the ecliptic ($7\overset{\circ}{.}005$) is larger than all other planets ("dwarf" planets excluded). Mercury's mean orbital velocity is 47.89 km/s. It is interesting and suggestive that the inclination is similar to the angle between the Sun's equatorial plane and the ecliptic, $7\overset{\circ}{.}25$.

Mercury also has the highest eccentricity of all the (major) planets, 0.2056. For JDN 2450480.5, the longitude of its ascending node, $\Omega = 48\overset{\circ}{.}3350$ and the longitude of perihelion was $\varpi = 77\overset{\circ}{.}4530$. As with the other planets, all of its elements vary with time. The closeness to the Sun produces an additional source of perturbation to Mercury's osculating orbits.

Due to its highly eccentric orbit, Mercury's orbit has been a testbed for gravitational theories. Misner et al. (1973, p. 1113) indicate that the line of apsides rotates forward (in the direction of orbital motion) at a rate of 55.9974(41) arc-sec/y (the parenthesis contains the uncertainty in units of the last decimal place: ±0.0041). Of this amount, 50.25645(50) arc-sec/y is due to "general precession" of the vernal equinox, a "contribution to the shift caused by the observer not being in an inertial frame far from the Sun."

Perturbations of other planets contribute 5.3154(68) arc-sec/y. The residual shift is 0.4256(94) arc-sec/y, which is of the order of the amount predicted by general relativity, but should include any effect of the solar oblateness (a non-spherical distribution of mass) as well.

Originally, Leverrier attributed this excess to a planet interior to Mercury, named *Vulcan*. This predicted planet has been proven fictitious; although there are asteroids in the inner portion of the solar system, their small masses and complex orbital movements make any of them unlikely to be responsible for the effect.

As noted in Chapters 3.8 and 5.3, this type of perturbation can, in principle, arise from an asphericity of the figure of the primary, in this case the Sun. Dicke and Goldenberg (1967) attempted to measure the oblateness of the Sun, which they used to evaluate the J_2 term in the expression for the solar potential. They found $J_2 \approx 3 \times 10^{-5}$. However, their experiment to demonstrate a systematic excess in the surface brightness of the Sun in the longitudinal direction compared to that in the latitudinal direction apparently did not take into account the changing brightness of the surface due to the transient active regions.

9.1.3 Mercury's Physical Properties

Mercury has a heavily cratered surface, and its features as viewed telescopically are only poorly resolved smudges. Between 1882 and 1890, Giovanni Schiaparelli observed the planet closely and reported a rotation period of 88^d. The proximity to the Sun was expected to result in tides on Mercury strong enough to cause a 1:1 spin–orbit lock, so this result was accepted for decades.

However, radio data from Mercury in 1962 suggested the presence of thermal radiation from the darkened disk of Mercury, an unexpected result if Mercury were indeed locked 1:1, because the "dark" side should be perpetually in darkness except for librational effects.

In 1965, R. Dyce and G. Pettingill used the Arecibo Radio Telescope to send radar pulses to Mercury and receive the return echo. The time delays and Doppler shifts were analyzed, resulting in a rotation period of 59^d compared to the 88^d orbital period. Modern values of these quantities are:

$$P_{\text{rotation}} = 58\overset{d}{.}6462$$

$$P_{\text{revolution}} = 87\overset{d}{.}969$$

Thus Mercury rotates three times on its axis for every two orbital revolutions. The geometry is shown in Figure 9.2.

The reason for Schiaparelli's conclusion is seen in Figure 9.3, based loosely on Fig. 4.9 of Consolmagno and Schaefer (1994). On average, an observer

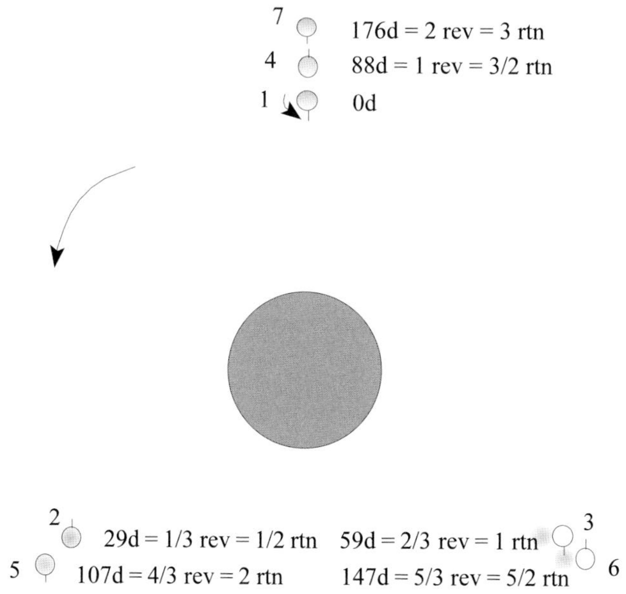

7
176d = 2 rev = 3 rtn
4
88d = 1 rev = 3/2 rtn
1
0d
2
29d = 1/3 rev = 1/2 rtn
59d = 2/3 rev = 1 rtn
3
5
107d = 4/3 rev = 2 rtn
147d = 5/3 rev = 5/2 rtn
6

Fig. 9.2. The 3:2 spin–orbit coupling of Mercury and consequences for a solar day at a particular location on Mercury (indicated by the very tall flagpole on the planet)

day 176
= 2P_rev
= 3P_rotn
day 264
= 3P_rev
day 88
= 1P_rev
day 352
= 4P_rev
day 0

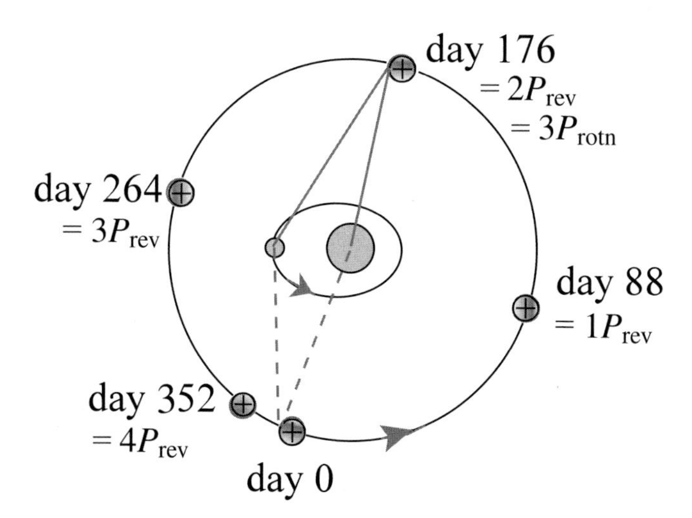

Fig. 9.3. Note that similar configurations of Mercury are seen from the Earth every two Mercurian sidereal years, which is equal to three Mercury rotations. Therefore during successive intervals of favorable maximum elongation (at aphelion), the same region of Mercury faces the Sun. Based loosely on Consolmagno and Schaefer (1994, Fig. 4.9)

on the Earth gets a relatively good look at the same region of Mercury only every fourth revolution (at every second 88^{d} interval the observer sees a different region of Mercury—but it too will be the same every fourth revolution). Consequently, because the aphelion apparitions of Mercury were more favorable for observation, Schiaparelli's data were aliased in such a way as to convince him, given the lock-in model for the Earth's moon, that he had the correct explanation. His data do indeed fit the 59^{d} rotation interval. It is seen that one of two longitude regions always faces the Sun at perihelion: longitude 0° and 180°. The *Caloris Basin* (Figure 9.4a), an aptly named multi-ringed plane, is near, although not centered on, longitude 180°.

This slow rotation explains its small amount of *flattening* or *oblateness*,

$$(\Re_{\rm eq} - \Re_{\rm pol})/\Re_{\rm eq} = 0.0 \tag{9.1}$$

The plane of the rotation is inclined by only $0\overset{\circ}{.}01$ to that of the orbit.

The radius of Mercury, from angular measurement via Mariner 10, is 2439.7 km.

Mercury has no moon, but in 1968, the asteroid Icarus approached within 16 million kilometers of Mercury, and subsequently space probes to Venus and Mercury were subject to perturbations from Mercury. These particular events and ongoing n-body numerical simulations of all the planets in the solar system produce more and more refined values of the mass of Mercury (as well as of the other planets). The 2005 Astronomical Almanac gives the value:

$$M = 0.33022 \times 10^{24}\,\mathrm{kg}$$

The combination of mass and volume provides the mean density:

$$<\rho> = 5430\,\mathrm{kg/m^3}$$

This compressed mean density is higher than that of any other planet except Earth.

Mercury has a low geometric albedo,[2] 0.106, and mean color indices: (B-V) = 0.93 and (U-B) = 0.41. These colors are significantly redder than sunlight. The lava flow areas are not, however, much darker than the highland areas (unlike the Moon where the mare basalts are much darker; see Taylor 1992, p. 193).

[2] The geometric albedo is defined in the AA as the ratio of the illumination of the planet at zero phase angle (i.e., the brightness as viewed from the light source) to that of a pure white Lambert plane surface of the same radius and position as the planet.

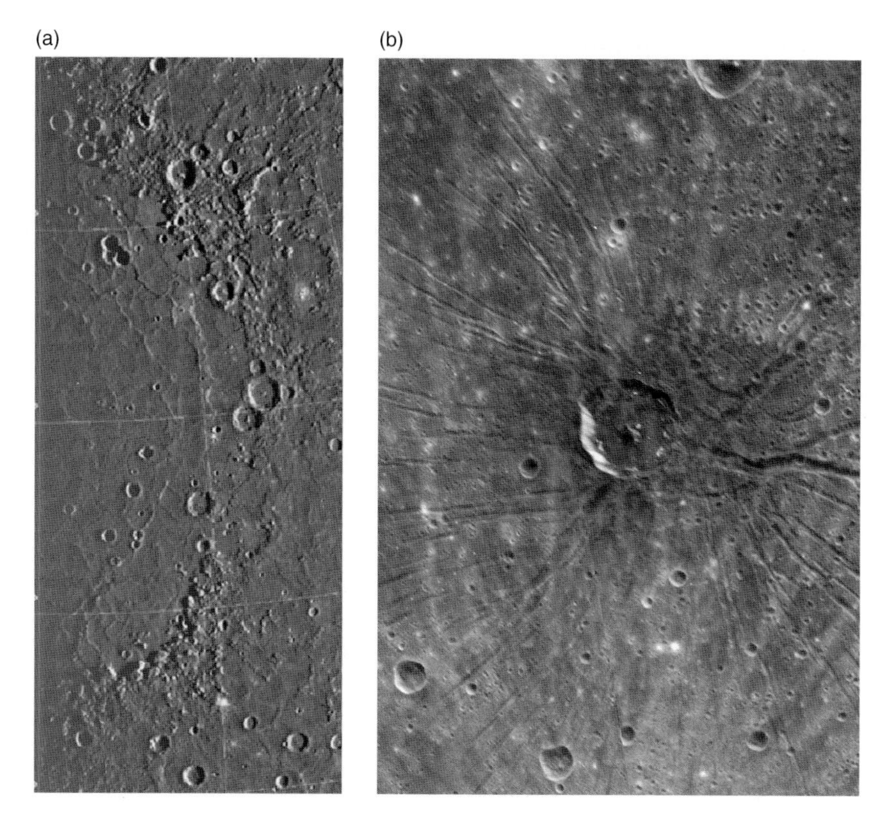

Fig. 9.4. **(a)** A portion of an annotated image of the *Caloris Basin*, on the left, as revealed by a Mariner 10 flyby of Mercury. JPL/NASA/Northwestern University image PIA024139. **(b)** A set of radiating troughs near the center of Caloris basin, photographed during the first Mercury flyby of the Messenger spacecraft, January, 2008, in an area that was in darkness during the Mariner 10 flybys (see Fig. 9.4(a)). The troughs, nicknamed "the spider," appear to have formed by extension and splitting of the floor material of the basin. A 40-km diameter crater lies close to the center of the pattern, but it is not clear if the troughs resulted from the impact or formed earlier, or perhaps a combination of the two. Image courtesy of NASA/Johns Hopkins University Applied Physics Laboratory/Carnegie Institution of Washington

Mercury's mean density suggests that there is a substantial amount of iron in the interior. Moreover, large amounts of this iron must be in metallic form, which has a closely packed crystal structure. Fayalite (Fe_2SiO_4, in the olivine group), the most iron-rich of the common silicates, has a density of only $4200\,kg/m^3$, insufficient to explain Mercury's density.

Yet the color and albedo of the *surface* suggest that very little metallic iron is present *there*. Compared to the reflectance spectra of lunar mare basalts, for example, there is only a very weak dip at 950 nm, where the lunar material shows strong absorption due to iron (Consolmagno and Schaefer 1994, p. 67).

9.1.4 Mercury's Origin

Mercury's history must include a way to concentrate its iron in the interior. Presumably the iron fell toward the center while the planet was in a molten state. The heating could come about from the heat of accretion, from radioactive material, or both, and might have occurred primordially, or, in a viscous interior, through a slower buildup of heat. In any case, the descent of iron in the interior must have contributed to Mercury's internal heat, and perhaps keeps it molten still.

Another clue is Mercury's magnetic field, detected by Mariner 10. It has a maximum value of 200 nT at 1.5$\mathfrak{R}$ on the sunward side.

The magnetic dipole field must have a value ~1% that of Earth.

A pure metallic iron core should have cooled rapidly, and, if the core is entirely solid, the existence of a magnetic field is "astonishing," as Lewis puts it. Taylor argues that the detection of a magnetic field indicates the existence of a molten core, possibly an outer core, as on the Earth (Taylor 1992, p. 191); he also suggests that this requires a large amount of FeS to be present in the melt region, with the combination having a lower melting point than metallic iron (this is called a *eutectic* mix, one in which two substances with high melting points when separate have a lower melting point when mixed together).

That the planet has substantially cooled is beyond doubt. Lobate *scarps* (height ~1 km; see Figure 9.4b) imply a contraction of ~2–3 km. A totally solidified core, would however, cause a decrease of ~17 km (Taylor 1992, p. 194).

Results of a recent study of the libration of Mercury (an oscillation of the body of the planet superimposed on its rotation) by Margot et al. (2007) gives the amplitude of the libration as 35.8(2) arc-seconds, and provides confirmation that Mercury has a molten core.

A sulphur content as little as 2–3% would allow the core (or outer core) to be molten, but it has generally been thought that volatiles, including sulphur, were driven outwards from this region of the early solar nebula before Mercury formed. A 2–3% sulphur content in Mercury would in fact obviate the need for such zoning. Even if this is correct, there is still the problem of the high mean density (compressed or uncompressed).

It has been argued that the high density is a result of a major impact early in the solar system history that removed much of Mercury's mantle (Benz et al. 1988, Wetherill 1988). Whether 30 or 500 initial bodies are used in simulations, the end result is 2–5 terrestrial planets in ~10^8 years. A major hit is certainly likely. Lissauer (1993, p. 153) cites a wide consensus in saying that the large range of rotational orientations in the solar system points to a major impacting period early in the solar system's history. The surface of this airless planet is heavily cratered, as Figures 9.5 and 9.6 demonstrate, and, because of this, its surface resembles the Moon more than it does any terrestrial planet. Indeed, aside from cratering impacts, the principal source of surface modification for both objects for aeons has been solar particles and radiation, meteorites of various sizes, and cosmic rays.

It is also possible for the volatiles simply to have evaporated in a high-temperature environment in the early nebula. The energy of accretion can be computed as follows. Assume that all mass falls from infinity (where it is at rest) in shells of mass $\mathrm{d}M$ on to a mass M. The expression becomes:

$$E = -G \int M(r)\, \mathrm{d}M(r)/r \tag{9.2}$$

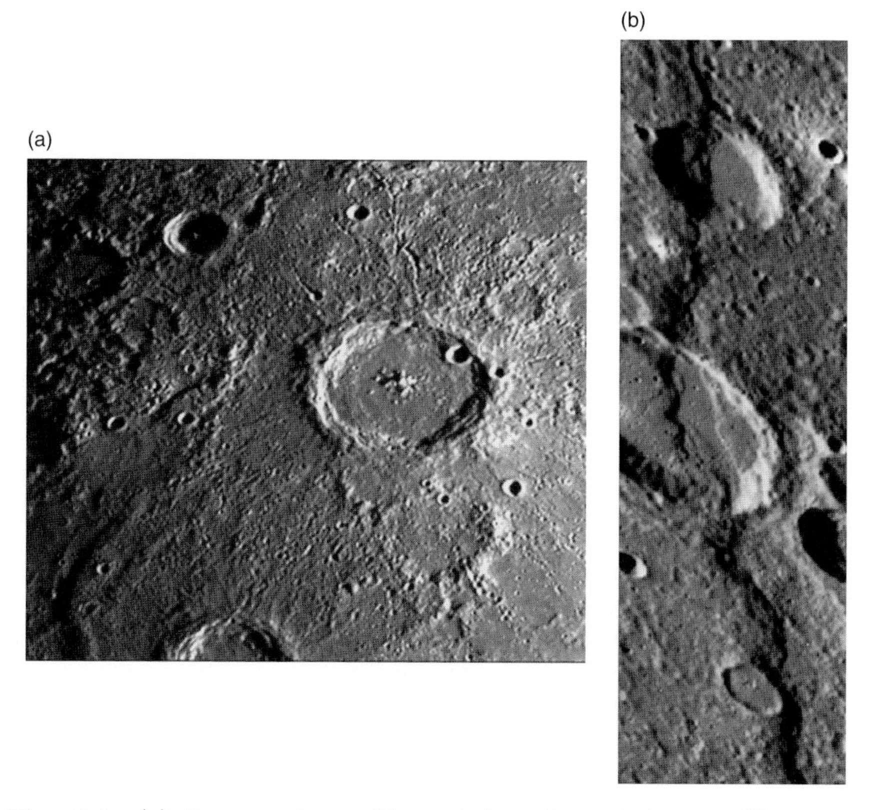

Fig. 9.5. (**a**) Large craters with central peaks and lava in-fill are not rare on Mercury. (**b**) The *Discovery Rupes*, a scarp seen at low Sun angle. JPL/NASA/Northwestern University images PIA02424 and PIA 02417, respectively

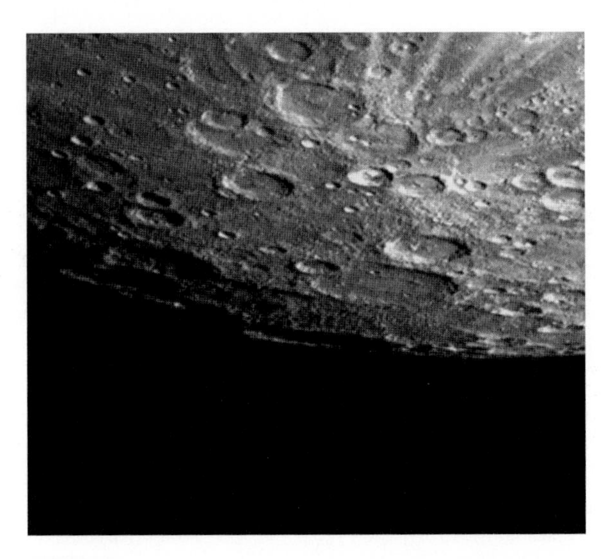

Fig. 9.6. The South Pole is in the crater with the illuminated far rim at the extreme bottom center of the image. Radar echoes suggest ice in the perpetually cold interiors of Mercury's polar craters. Bright crater ejecta rays are seen at the top of the image. Mariner 10 image (NASA/JPL/NorthwesternUniversity, PIA02941/02415)

where the integration is taken from 0 to M. The result is:

$$E = -[(GM^2/(2R)] \tag{9.3}$$

Substituting $\rho = M/[(4\pi/3)R^3]$, and allowing for non-uniform density, leads to

$$E = -k\,(16\pi^2/9)\,G\rho^2 R^5 \tag{9.4}$$

where the quantity k varies from 3/5 for a uniform composition to 3/2 for main sequence stars. This expression is found in Lewis (1995, p. 398), with $k = 1$. Equation (9.4) with Mercury's present values (and $k = 1$), gives 3.0×10^{30} J. Lewis allows for 2/3 Fe and 1/3 magnesium silicates, and computes a temperature increase of 15,000 K, certainly sufficient to evaporate most gases! The student can pursue the matter further!

Currently, Mercury shows signs of accreted water on its surface, presumably from low-speed comet impacts. Note that the poles get no direct illumination and T must be below the freezing point of water in the perpetually shaded interiors of deep polar craters (see Figure 9.6). Radar echoes indicating high albedos from both polar regions have been detected, implying the presence of ice.

The mean solar day length is found from the relative angular rate formula:

$$\omega_{\rm msd} = \omega_{\rm rtn} - \omega_{\rm rev} \tag{9.5}$$

$$\therefore\ 2\pi/P_{\rm msd} = 2\pi/P_{\rm rtn} - 2\pi/P_{\rm rev} \tag{9.6}$$

from which we find

$$P_{\rm msd} = 179{\rm d}$$

For an object of Mercury's periods of revolution and rotation, the Sun would be above the equatorial horizon for half of this interval, or 89d. However, at perihelion Mercury's large eccentricity causes it to sweep around the Sun faster than the rotation ($\omega_{\rm rev} > \omega_{\rm rtn}$). Therefore, the Sun on Mercury would normally seen to move from east to west, then, near perihelion, the following sequence would be observed:

Sun becomes stationary
Sun moves eastward
Sun becomes stationary
Then the Sun moves westward again, repeating the cycle

The visibility of the Sun in the sky for other latitudes on Mercury differs only slightly from that at the equator in ways which are better illustrated for planets with much larger differences between their rotational and revolution axes, so we leave this description to the interested reader! Further discussion of the nature of Mercury, including such topics as the surface reflectance, and the crater size/frequency distribution, compared to those of the Moon, can be found in Strom and Sprague (2003).

9.2 Venus

9.2.1 Visibility and General Properties

Venus sometimes appears as the brightest object in the sky after the Sun and Moon, achieving −3.3 to −4.4 V magnitude at ~39° elongation (not quite at maximum elongation: 48°). As either a morning or an evening star, it commands attention.

The planet was considered a manifestation of a god in many past cultures (e.g., as Ishtar/Inanna in the Middle East, Hesperus in its evening star manifestation and Phosphorus [Lucifer, bearer of light] in the morning, further West). In Mesoamerica, it was considered a fearsome god, and the spilling of blood was required to placate it. For extensive discussion of the perception of Venus from culture to culture, and its configurations in the sky, see Kelley and Milone (2005).

The bulk properties of the planet are summarized in Table 3.1. Its orbital eccentricity (0.007) is smaller than that of the Earth, so its orbit is more circular. The orbit is also close to the ecliptic plane: $i = 3.39°$.

Its period of revolution is 224.7d, so that its synodic period is 1y 219d = 1.599y. Transits of Venus are observable when inferior conjunctions occur near a node passage. The most recent transit occurred on June 8, 2004 and before that, on December 6, 1882; the next three will occur on June 6, 2012, December 11, 2117, and December 8, 2125. Note the pairing of transits in short intervals, and the change of node across a much larger interval. Such events were sought to determine the scale of the solar system, as soon as ephemerides could be accurately computed—following Johannes Kepler. The distance rotated through by the Earth in the course of the transit can be used as a baseline for a single observer (Method 1 in Figure 9.7). Alternatively, the known separation of two observatories on the Earth can provide the baseline of a triangle, the apex of which is at Venus (e.g., Method 2, in Figure 9.7). The relative timings of the immersion and emersion and/or the exact placement of chords across the Sun provide the necessary data. With either method the

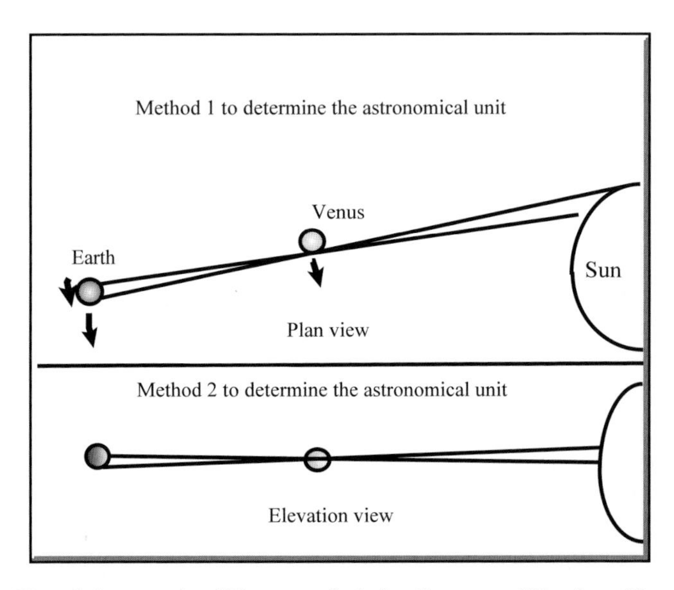

Fig. 9.7. Use of the transit of Venus to find the distance of Earth to Venus in kms, and thus the astronomical unit

Earth–Venus distance can be found in kilometers. Setting this distance equal to the fraction of the astronomical unit expected from celestial mechanics, the value of the astronomical unit is found:

$$r\,(\mathrm{km}) = f\ a\,(\mathrm{km}) \tag{9.7}$$

where r is the determined distance, a, the astronomical unit and f, the fraction of the au represented by the distance.

Venus, like Mercury, has no natural satellite. From n-body integration work, however, its mass is found to be 0.815 that of Earth. By direct measurement before and after inferior conjunction, its radius is 6052 km, so that its mean density is $5200\,\mathrm{kg\,m^{-3}}$. This implies a large iron core (see below). Yet, Venus has no detectable magnetic field unrelated to the solar wind. This may be due to Venus' slow rotation, and/or the absence of a *liquid* core.

The mean gravitational acceleration on the surface of Venus at the equator is $8.96\,\mathrm{m/s^2}$, making it the most Earth-like planet in terms of weight on the surface. Aside from the physical measurements, however, the environment on the surface of Venus is unmatched on Earth, except maybe *in* active volcanoes!

As for Mercury, the rotation period has not been easy to measure, but here the dense clouds that veil the planet are the cause. In 1890, Giovanni Schiaparelli (1835–1910) suggested that $P_{\mathrm{rotation}} = P_{\mathrm{rev}}$, and Percival Lowell (1855–1916) concurred in 1911. In 1921, Edward C. Pickering (1846–1919) produced a value of 68^{h} about an axis in the plane of the orbit; W. H. Steavenson in 1924 confirmed this axis, but found a period of 8^{d}. In 1927, Frank E. Ross (1928) proved from UV-emulsion photography carried out with the 60- and 100-in telescopes on Mt Wilson that so short a period as 68^{h} could not satisfy his data (references to earlier work are cited in his paper), but he was unable to find a period with much precision because of the elongated nature of the markings and their changing appearance from day to day, and settled on a value $\sim 30^{\mathrm{d}}$. In 1956, J. D. Kraus claimed that radio signatures had yielded a period of $22^{\mathrm{h}}\ 17^{\mathrm{m}} \pm 10^{\mathrm{m}}$. Radar determinations in 1962 finally produced an accurate result: $-243^{\mathrm{d}} = -0.665^{\mathrm{y}} = 1.50\,P_{\mathrm{rev}}$. Because the synodic period of Venus is 1.60^{y}, Venus is close to an orbital lock-in with Earth (Chapter 3.7.3).

The conditions on the surface of Venus are beyond "harsh!" The Sun would be perceived only dimly through the super-dense cloud decks, and the greenhouse effect produces a surface temperature of $\sim 750\,\mathrm{K}$, under a pressure of ~ 90 bar at the surface. The atmospheres of Venus and the other terrestrial planets will be compared in Milone & Wilson (2008, Chapter 10), so we defer further discussion of Venus' atmosphere until then.

Next we describe the features on Venus, known only through radar observations and mainly from the Magellan mission of the 1990s. Other data will be discussed in the sections to follow.

9.2.2 Types of Surface

9.2.2.1 Division in Terms of Elevation

Highlands:	8% of Venus' surface area
Rolling plains:	65%
Lowlands:	27%

Venus has relatively little relief compared to the Earth or Mars, and the frequency distribution of its features' altitudes is unimodal (i.e., one peak near ∼0 altitude, compared to Earth's distribution which is bimodal with much greater variation); its highest mountain, Maxwell Montes, however, is higher than Mt Everest. See Figures 9.8 and 9.9a and b, and, for further discussion, Head and Basilevsky (1999).

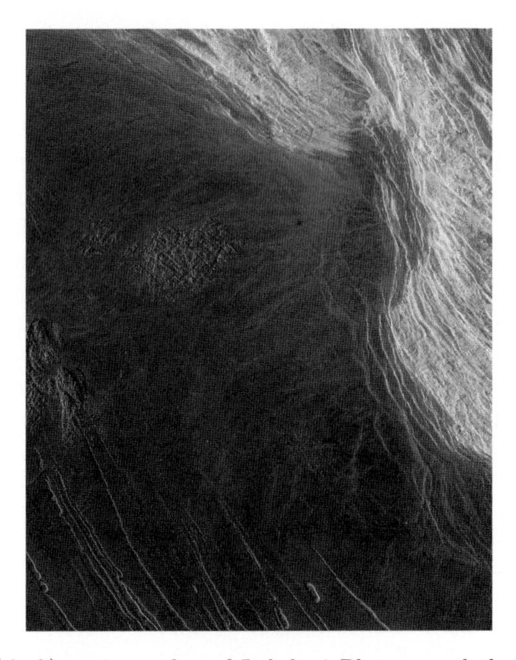

Fig. 9.8. The (*dark*) eastern edge of Lakshmi Planum and the western edge of Maxwell Montes, at 11.5 km, the highest region on Venus. At left center are tesserae, jumbled terrain due to intersecting graben. Other graben at bottom. A Magellan NASA/JPL image, PIA00241

Fig. 9.9a. A false color contour map of the northern hemisphere of Venus, based on Magellan radar data. Lakshmi Planum and Maxwell Montes are seen just below the center. NASA/JPL image PIA00007

Fig. 9.9b. A false-color contour map of the southern hemisphere of Venus, constructed from Magellan radar data. The extensive bands on the lower right limb are features of an equatorial rift zone. NASA/JPL image PIA00008

9.2.2.2 Division in Terms of Geologic Origin

Volcanic units (lava plains):	>70%
Highly deformed tectonic[3] units:	25%

9.2.3 Major Geologic Features of Venus

9.2.3.1 Volcanic Origin

Lava plains: These are basaltic flows with some basaltic debris (wind-blown deposits and/or volcanic outthrows).

Circular rises: These features are 50–300 km diameter, usually <1 km high, and resemble the Earth's shield volcanoes.

Volcanic domes: These objects range from a few km to 15–20 km in diameter, often with a summit crater.

9.2.3.2 Tectonic or Tectonic + Volcanic Origin

Tesserae: A very common type of upland terrain with an appearance reminiscent of a parquet floor. ("Tessera" is from Greek, meaning "tile.") They are dominated by densely-packed systems of ridges and grooves crossing each other in an orthogonal, diagonal, chevron-like, and/or chaotic manner. The ridges are typically 5–20 km apart and several hundred meters in height.

Mountain belts: These are parallel ridges 5–20 km apart and several hundred meters in height, at elevations of one to several kilometres above the surrounding plain. Examples: The Akna, Danu, Frejya, and Maxwell Montes around Lakshmi Planum in Ishtar Terra (Figure 9.8).
The Frejya and Akna Montes are long and linear, with synclines and anticlines, thrust faults, and strike–slip faults. The Maxwell Montes seem to be similar but have also been compressed parallel to the ridges with movement along faults perpendicular to the ridges, making the mountain area more equi-dimensional.
The mountain-belts are usually interpreted as being of compressional origin, like terrestrial folded mountains.

[3] The word "tectonics" refers to any geologic process which involves the movement of solid rock. Large-scale tectonics in the crust is generally caused by processes in the underlying mantle. In plate tectonics on the Earth, solid lithospheric plates slide around on the surface in response to convection in the mantle. Plate tectonics does not seem to occur on Venus, but other tectonic processes do.

Ridge belts: Belts of ridges and grooves up to several hundred kilometers wide and many hundreds of kilometers long. The ridges are similar in height and spacing to the mountain belt ridges, but are usually not elevated above the surrounding plains by more than a few hundred meters, and sometimes are even located in shallow troughs.

The ridge belts are interpreted as compressional features, sometimes with later extensional deformation.

Regiones: Large plains-like uplands with grooves and scarps along their crests, and volcanism. They result from rifting associated with tectonic updoming. Examples: Beta Regio, an elevated rift zone with a large shield volcano (Theia Mons) that has partly flooded the rift valley.

Coronae (ovoids): Circular or elliptical structures with 150–1000 km diameter. They are often surrounded by rings of concentric fractures or parallel ridges and grooves. The area inside the ring is often highly fractured and elevated above the surrounding plains. They may be caused by doming over isolated regions of upwelling in the mantle (hot spots).

The largest coronae, such as Artemis Corona (Figure 9.10) on the south side of Aphrodite Terra, resemble terrestrial subduction zones: both are arc-shaped with similar radii of curvature (300 km or more)

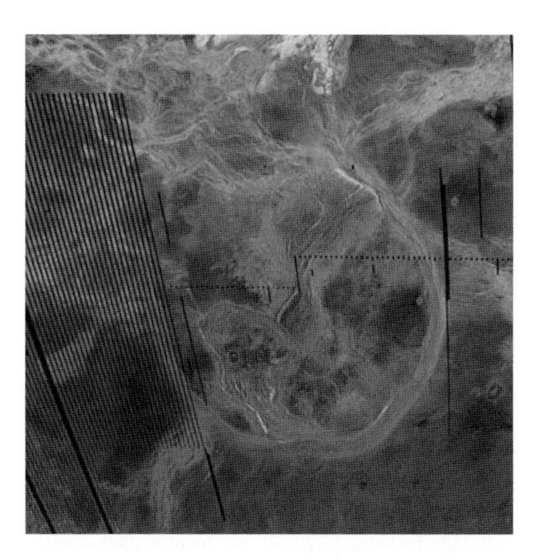

Fig. 9.10. Artemis Corona, one of the largest such volcanic constructs found on Venus. The dark linear striations are artifacts of the recording and transmission process from the Magellan spacecraft. NASA/JPL image, PIA00101

and both have an outer rise with a deep trench on its interior side, and another, lower, ridge interior to the trench.

Lakshmi Planum is surrounded by the Frejya, Akna, Danu, and Maxwell Montes and may be a "megacorona." A cross-section through the outer part of Latona Corona on Venus is shown in Figure 9.11, from Sandwell and Schubert (1992), compared to a cross-section through the Sandwich subduction trench on Earth. Both trench arcs have radii of curvature ~340 km.

Arachnoids: The term *arachnoid* is from the Greek word for "spider." They range from 50 to 200 km in diameter. Figure 9.12 portrays an arachnoid from the Fortuna region.

Arachnoids are similar to coronae but smaller and with a system of radial features (forming the "legs" of the "spider"). Their origin is unknown, but may be similar to that of the coronae and may, in fact, represent an early stage in the evolution of coronae.

A possible model for the evolution of a large corona is shown in Figure 9.13, adapted from Sandwell and Schubert (1992).

9.2.3.3 Comparative Plate Tectonics Earth Vs. Venus

Plate tectonics requires three conditions:

1. Rigid lithospheric plates
2. Creation of new lithosphere along one boundary of the plate
3. Subduction of lithosphere along another boundary of the plate

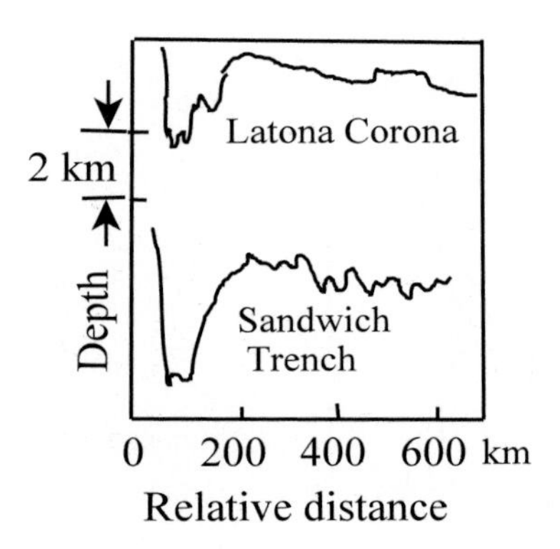

Fig. 9.11. Profiles of the Latona Corona on Venus and a subduction are feature on the Earth: the Sandwich Trench in the Pacific Ocean. After Sandwell and Schubert (1992, Fig. 2); adapted with permission of AAAS

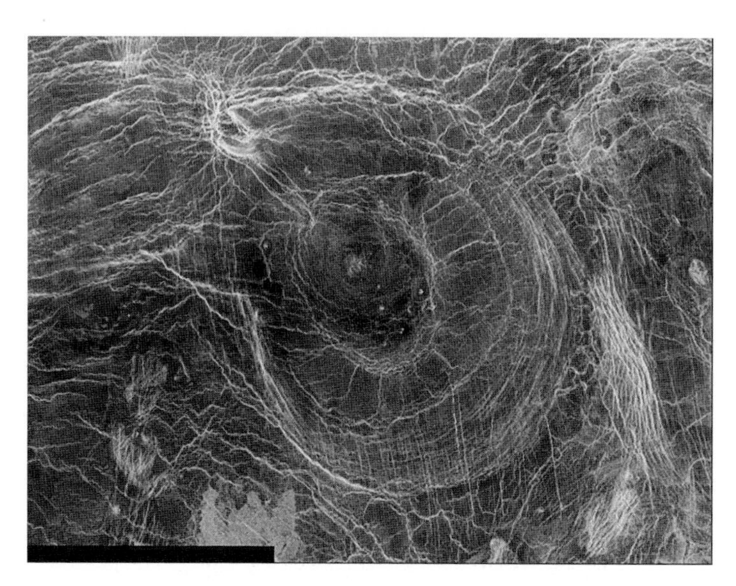

Fig. 9.12. An arachnoid in Fortuna. The radial cracks may be associated with upwelling magma within the dome. Magellan NASA/JPL image

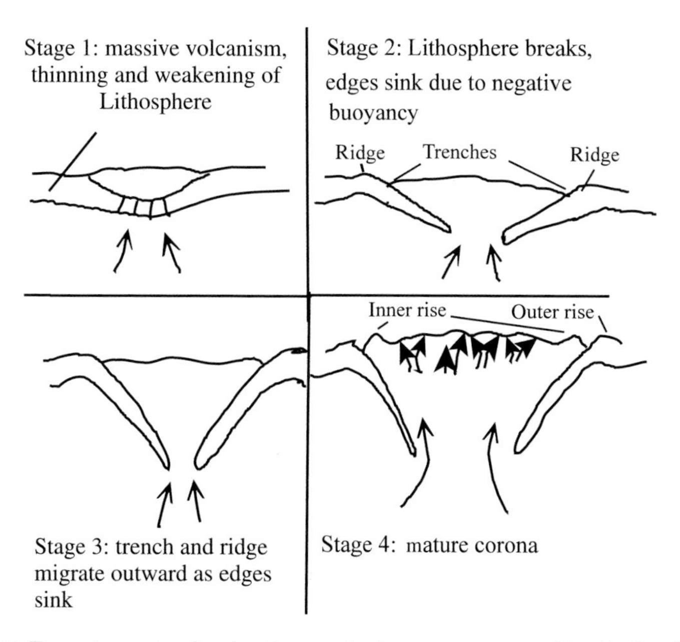

Fig. 9.13. Four stages in the development of a corona, according to Sandwell and Schubert (1992), adapted from their Fig. 6, with permission of the AAAS. The arrows represent upwelling mantle material

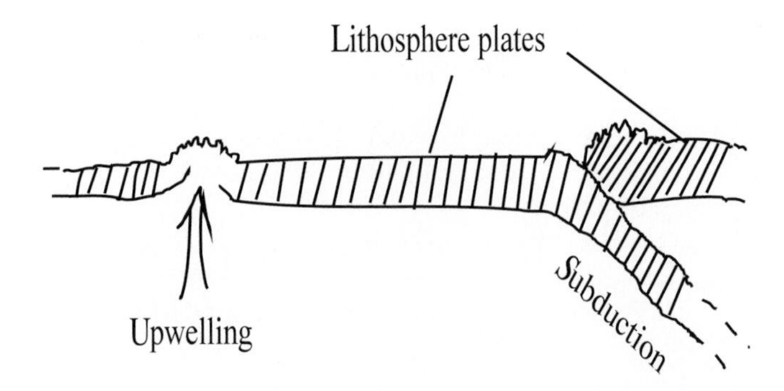

Fig. 9.14. Seafloor spreading and subduction of the lithosphere on the Earth

On Venus, the lithosphere is not as rigid as on Earth because of the ~450 K higher surface temperature. The lithosphere may therefore deform without being subducted, inhibiting global plate tectonics. Figure 9.14 illustrates plate tectonics on Earth from the viewpoint of the processes acting on a plate.

Five types of forces are involved in plate tectonics; they are described below and compared to the expected forces on Venus, using the numbers in Figure 9.15.

(1) Ridge Push Forces acting along a mid-ocean ridge push the two plates apart:

A. The material rising from the mantle physically pushes the plates apart.

B. The ridge is elevated, so the weight of the plates makes them slide down each side of the ridge, pushing outward on the two plates.

On Earth, the force due to the rising material is larger than the sliding force by about an order of magnitude.

Ridge push may be important in initiating subduction by forcing the far edge of a plate to thrust under an opposing plate.

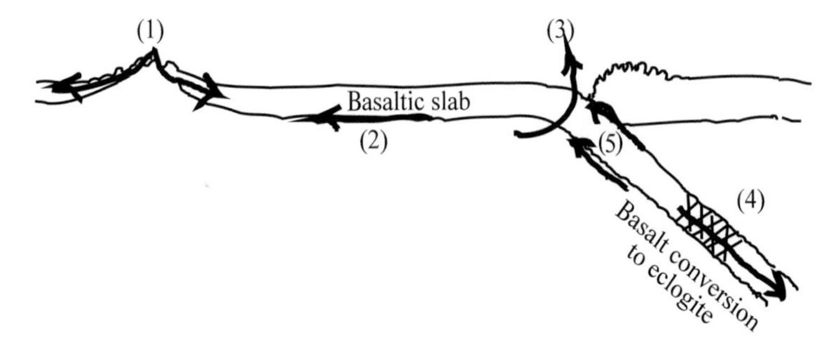

Fig. 9.15. The five forces of tectonics illustrated

On Venus, the lithosphere is hotter than on Earth because of the ~450 K higher surface temperature. The implications are that:

The lithosphere is softer and less able to support high elevations.
A "mid-ocean ridge" on Venus would be expected to be only about 40% as high as on Earth.
It would also be less dense because of thermal expansion of the rock.

Ridge push is therefore expected to be much less effective on Venus than on Earth.

(2) Resistance to Movement over the Underlying Mantle For this to be important, the flow speed of the underlying mantle must be *less* than the speed of the plate. (If the mantle were moving faster, it would provide a *driving* force.)

A larger resistance reduces the size of a plate which is capable of sliding as a unit over the mantle.

On Earth, resistance to plate motion is greatly reduced by the partially-molten low velocity zone (LVZ) below the lithosphere.

On Venus, there may not be an LVZ. If there is not, then the much larger resistance to plate motion would greatly reduce the maximum size of plate possible.

(3) Resistance to Bending The lithospheric plate is solid and resists being bent at the subduction zone. The resistance is larger for a thicker plate, but reduced if there is an overlying layer (e.g., ocean water on the Earth).

On Venus, we expect a thinner lithosphere but no oceans, so the bending resistance may be very similar on Venus to that on the Earth.

(4) Slab Pull The lithospheric plate (density ρ_L) rests on and descends into the mantle below it (density ρ_M).

The descending slab feels two forces (Figure 9.16):

1. The weight downward
2. The upward buoyancy forces (equal to the weight of an equal volume of displaced mantle)

The net force (the "buoyancy") depends on the difference in density between the mantle and the descending slab, $\rho_M - \rho_L$.

The buoyancy is positive or negative depending on whether this net force is upward or downward:

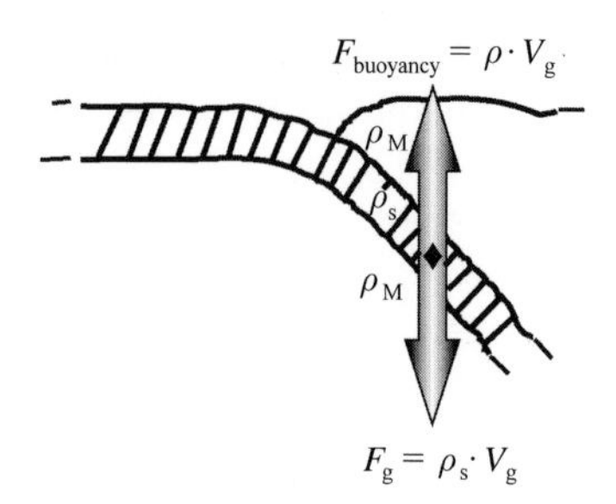

Fig. 9.16. Slab pull: forces on a descending plate

Case 1: $\rho_L < \rho_M$ (positive buoyancy). The slab is less dense than the surrounding material and floats on the mantle.
Case 2: $\rho_L > \rho_M$ (negative buoyancy). The slab is denser than the surrounding material and can sink into the mantle.

On Earth, the descending slab is cooler than the mantle it descends into because it has spent 200 million years or so on the surface. As it descends, it is warmed by conduction, but this is a slow process; the slab remains cooler than its surroundings to a considerable depth. Being cooler, it tends to be denser and therefore negatively buoyant. The cooler temperature also causes the basalt–eclogite (at about 50–100 km) and the olivine–spinel (at about 300–400 km) phase changes to occur sooner, increasing the negative buoyancy.

The slab pull force appears to be about 10 times larger than the ridge push force, making it the dominant force sustaining subduction once subduction has begun.

On Venus, under a higher temperature,

The lithosphere is less dense due to thermal expansion
The transition from basalt to eclogite is deeper, so the lithosphere remains more buoyant to greater depths

Slab pull may therefore be much less effective on Venus due to the greater buoyancy of the lithosphere.

(5) Shear Forces on the Descending Slab These arise from:

Shear between the slab and the opposing plate
Shear between the slab and the underlying mantle

Shear between the slab and the opposing plate is greatly reduced by the presence of water (a major factor on Earth).

On Earth, the shear between the slab and the underlying mantle is apparently zero! This is possibly due to water.

On Venus, there is no water. Shear forces may therefore strongly inhibit subduction.

9.2.4 Magellan Spacecraft Results

No major plate tectonic features have been seen. This includes:

No interconnected spreading centers (mid-ocean ridges)
Subduction trenches appear to be limited to individual hot spots (coronae) and do not lie along interconnecting plate boundaries

The primary resurfacing processes are volcanism and tectonics but still less than on Earth. The evidence comes from impact craters :

- The observed cratering ages range from a few million years to ~1 Gy, compared with about 200 million years maximum age for Earth's ocean floors. Figure 9.17 provides an example of an impact crater in the process of being resurfaced.
- A few large areas have no craters, e.g., $5,000,000\,\mathrm{km}^2$ in the Sappho region (between Beta Regio and Aphrodite Terra). This implies that volcanic resurfacing is locally efficient but episodic.

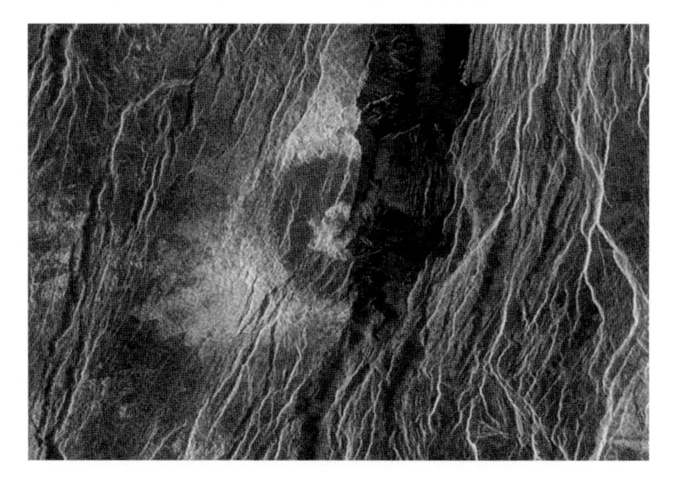

Fig. 9.17. The Summerville impact crater, 37 km diameter, half obliterated by faults in a rift area of Beta Regio. Part of the central peak region can be seen sliding into the chasm. A Magellan NASA/JPL image, PIA00100

Fig. 9.18. Wind-blown deposition streaks at the Adivar impact crater on Venus as seen by Magellan. NASA/JPL image, PIA00083

- Erosion does not appear to be very important on Venus because there is no water; and few wind-blown deposits have been seen. There are some, however. For example, see Figures 9.18 and 9.19.

9.2.5 Crust

Three types of crust can be defined for terrestrial planets

1. Primary crust: solidification on cooling following accretional heating. Example: the lunar highlands.
2. Secondary crust: following partial melting of the mantle; this produces basaltic crust; Examples: lunar maria, Earth's oceanic crust.

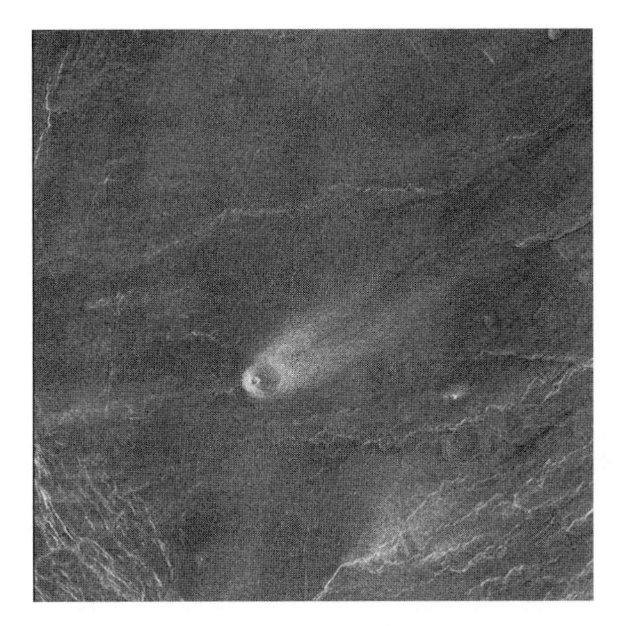

Fig. 9.19. A 35-km long deposition streak of radar-bright material on a volcano in the Parga Chasma region of Venus. A Magellan NASA/JPL image, PIA00243

3. Tertiary crust: following remelting of the secondary crust at the base of a thick crust or on subduction or foundering of the crust into the mantle; this produces granitic crust. Example: Earth's continents.

On Venus, the low crater density, total lack of impact basins >160 km diameter, and obvious resurfacing indicate that no primary crust is left.

9.2.6 Tectonics on Venus

In this summary we are indebted to the comments of Kerr (1991).

For the Aphrodite Terra and Ovda Regio:

1. Ovda Regio (Figure 9.20) is part of western Aphrodite Terra.
2. Two leading alternatives have been proposed for the observations:

   (a) *Blob tectonics*

   Hot material in the mantle rises in individual blobs toward the surface.

   The mantle material between it and the lithosphere is squeezed upward and outward ahead of the blob.

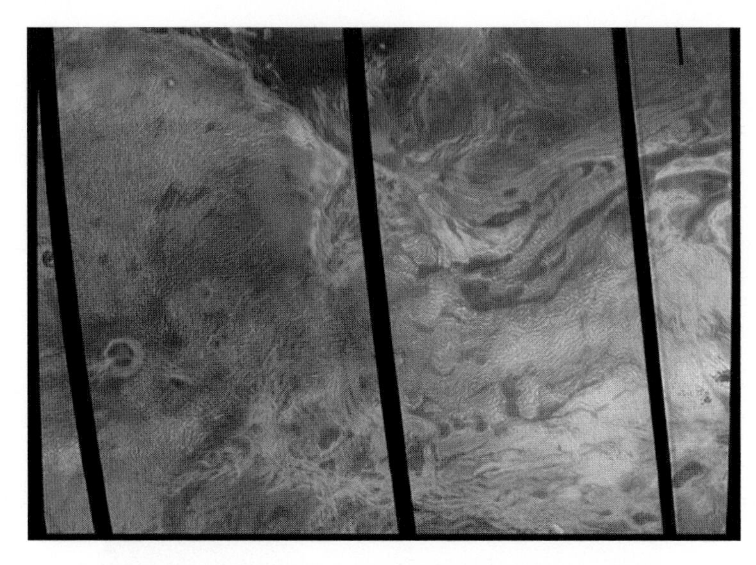

Fig. 9.20. Ovda Regio, a highland region of Venus in western Aphrodite. It rises ∼4 km above the low-lying plains. It is a region of mountain belts, domes, ridges, and valleys. A Magellan NASA/JPL image, PIA00146

This pushes the lithosphere, including the crust, up into a dome.

The blob then flattens against the bottom of the dome and spews magma onto the surface to form a volcanic hot spot, as in Figure 9.21.

In this model, four such hot spots make up western Aphrodite Terra, including Ovda Regio, with each hot spot in a different stage of a blob's life cycle.

(b) *Convergence model*

The mantle drags the lithosphere around with it.

Where the mantle converges and sinks, the crust bunches up without being subducted. (See Figure 9.22.)

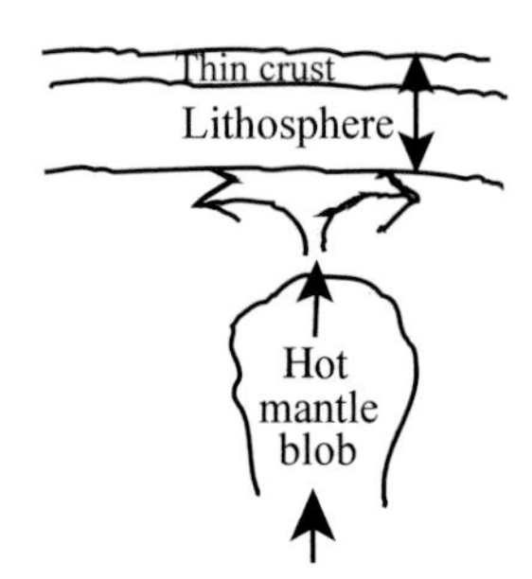

Fig. 9.21. Model 1: blob-tectionics

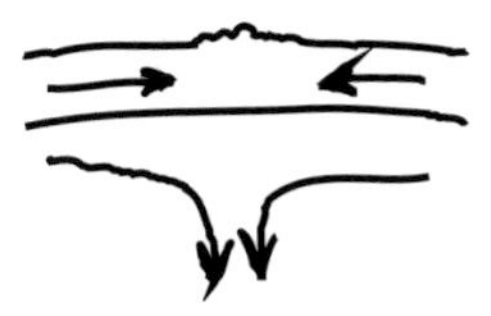

Fig. 9.22. Model 2: convergence

9.2.7 Surface Information from the Soviet Landers

Much of this material is summarized in Fegley et al. (1997).

9.2.7.1 Terrain at the Lander Sites

Venera 8: Domes and lava flows in plains east of Navka Planitia.

Venera 9: Steep (15–20°) slope of a hill in Beta Regio, densely covered by ∼10-cm-size plate-like rock fragments and some loose soil in the depressions between.

Venera 10, 13, 14: Plains, surface dominated by low-standing flat-topped outcrops of bedded rocks with variable amounts of loose soil material in local lows. Venera 10 was in lowlands near the southeastern edge of Beta Regio. Venera 14 landed on the flank of a volcano in the south of Navka Planitia.

9.2.7.2 Chemical Composition This was sampled at seven locations: Veneras 8, 9, 10, 13, 14, and Vegas 1 and 2 sites.

All except Venera 9 were in plains near the equator. Venera 9 was on a slope near Rhea Mons (in Beta Regio).

The measured surface temperature was 748 K (475°C) and the pressure, 90 bars.

The wind speed was less than 1 m/s (3.6 km/h). This wind was strong enough to gradually decrease (over the space of about an hour) the size of a clump of soil which had fallen onto the supporting ring of one of the landers during the landing.

Most of the data indicate that the material around the landers, including bedded rocks, is porous and friable, with densities in the range ~ 1400–$1500\,\mathrm{kg/m^3}$, except for Venera 10 site where the density was around $2800\,\mathrm{kg/m^3}$.

9.2.7.3 Techniques to Measure Composition Two techniques were used:

1. *γ-ray spectroscopy*. This was carried out by Veneras 8, 9, and 10, and Vegas 1 and 2 (which landed in/near Rusalka Planitia and carried no TV

imagers). It gave the abundances of K, Th, U in the surface layer under the lander.

2. *X-ray fluorescence.* This was carried out by Veneras 13 and 14 and Vega 2. A centimeter-size drill sampled beneath each lander. It gave the abundances of Si, Ti, Al, Fe, Mn, Mg, Ca, K, S, and Cl.

9.2.7.4 Composition at the Soviet Lander Sites These came from five sites (Veneras 9, 10, 14, Vegas 1 and 2).

The rock was found to be similar to terrestrial *tholeiitic basalt.* Tholeiitic basalt is more silica-rich than other basalts. On Earth it is produced along the mid-ocean ridges and makes up the oceanic crust.

This indicates that the rock represents secondary crust, that is, crust derived directly from material rising from the mantle.

From the individual landers:

Venera 13. This probe landed in the Navka Planitia in the eastern part of the Phoebe Regio. Rocks were found to be similar to terrestrial subalkaline basalts, which are found in rift areas in the Mediterranean region.

Venera 8. Rocks were similar to terrestrial alkaline basalts, and resemble the granitic continental crust on Earth. This was interpreted by some Soviet scientists in the 1980s as indicating ancient crust created by the early differentiation of Venus as it cooled, in a manner similar to Earth. This crust would then have become overlain at a later date by the younger tholeiitic basalts rising from the mantle to form the volcanic regions and the large volcanic plains.

However, the Magellan images (Figure 9.23) show the Venera 8 landing site to be in a region of mottled plains with a complex of flows interpreted

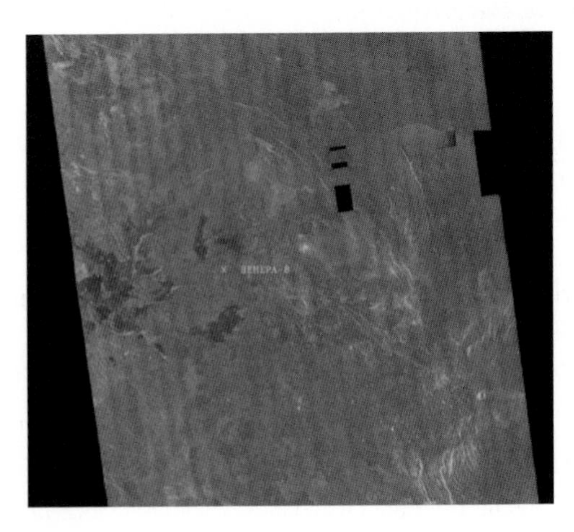

Fig. 9.23. The Venera 8 landing site in the Navka region on Venus. Magellan image, NASA/JPL PIA00460

as basaltic lava flows, related to NW-trending fractures in the plains. Nearby is a pancake dome, 22×25 km in extent, made up of a shallow inner depression 10–12 km across surrounded by a raised annulus 5–6 km wide with steep slopes. The inner depression has several shallow, rimless pits. The dome is surrounded with a concentric fracture pattern extending out about 10–15 km. The dome is similar in appearance to rhyolite, dacite, and andesite domes on Earth (rhyolite is the volcanic equivalent of granite). Venera 8 may thus have measured rocks associated with this dome, rather than ancient crustal rock. This is supported by the young age (<1 Ga) of all surface features seen so far by either Magellan or Veneras 15/16.

9.2.8 Interior Heat Budget of Venus

Surface U, Th, K are similar to Earth, implying similar rates of radiogenic heat generation.

If Venus loses heat at the same rate as Earth per unit mass, then the heat flux from the surface is about $70\,\mathrm{mW/m^2} \pm 30\%$ (the uncertainty is due to possible differences in potassium abundance and differences in secular cooling.

Secular cooling is time-dependent cooling from an earlier, hotter state, e.g., from accretional heating during planetary formation.

On Earth, the secular cooling rate is about 100 K per Ga in the mantle and may contribute up to 50% of Earth's heat budget.

Secular cooling may be slower on Venus because the conductive heat loss through Venus' lithosphere is inefficient compared to plate tectonics on the Earth. This would reduce the heat flow and may result in a hotter interior.

The extensive lava flows on the surface of Venus may be due partially to the high temperature environment and partially to dissolved gases within it, which may make the lava more fluid. On Earth, steam and other dissolved gases are released from the molten material at the surface, but on Venus, the 90 bars of surface pressure are sufficient to keep these gases in solution, resulting in a smoother flow than is found on Earth (Head and Basilevsky 1999). On the other hand, in another interpretation the "pancake domes" have been attributed to thick,viscous lava, which formed a cap that blocked further lava flow. There is some evidence, however, that at least some flows on Venus are less viscous than Earth lavas (note the arrow-marked, very long flow channel in Figure 9.24).

9.2.9 Methods of Heat Loss Through Planetary Lithospheres

The three ways in which heat can escape the interior are summarized in Figure 9.25.

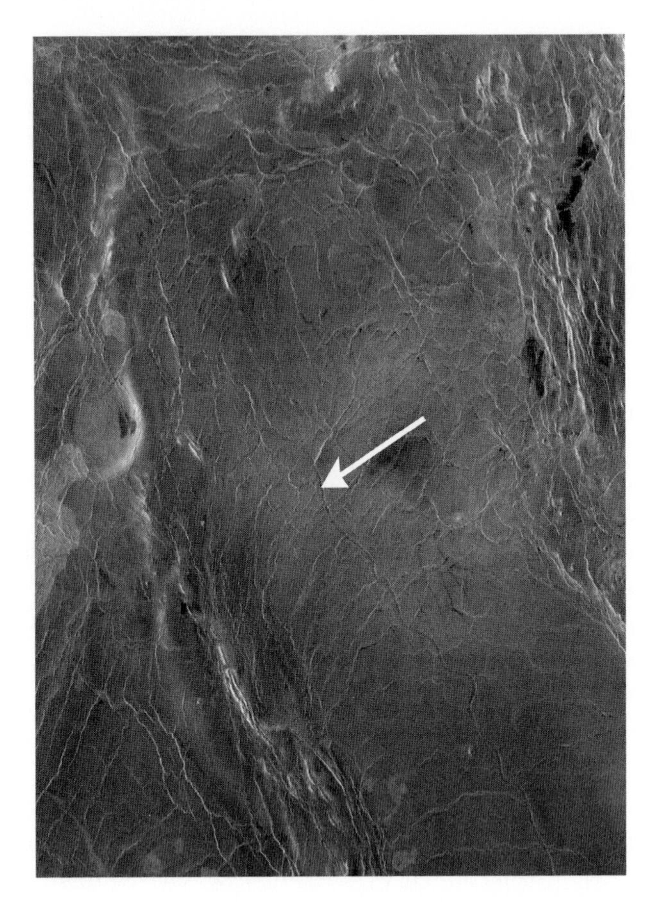

Fig. 9.24. Methods of heat transport to the surface for rocky solar system bodies. A 600 km portion of the longest channel detected on Venus. Discovered by the orbiters during the Venera 15–16 mission, it is more than 7000 km long and ~1.8 km wide. A common type of feature on Venus, it is thought to be a lava channel, evidence that at least some types of lava on Venus are less viscous than Earth lavas. Magellan/NASA/JPL image PIA00245, with added marker

9.2.9.1 Advection Latin *ad* (to) + *vectum* (carried): The heat is carried to the surface by material transport such as volcanoes, lava flows, and subsurface intrusions of magma.

This is the dominant heat loss mechanism for Io (Jupiter's innermost Galilean satellite; see Chapter 13)

9.2.9.2 Lithospheric Conduction This is the dominant heat loss mechanism on planets for which plate tectonics, volcanoes and lava flows are not currently important. It is the sole heat loss mechanism for the Moon and Mercury, and the dominant mechanism for Mars and Venus.

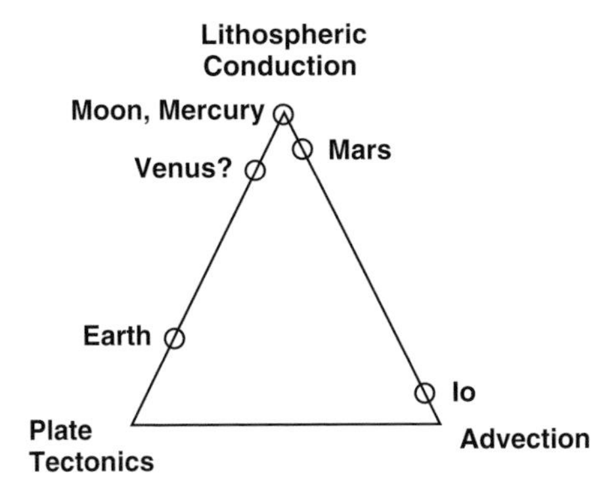

Fig. 9.25. Methods of heat transport to the surface for rocky solar system bodies. The importance of each process for the terrestrial planets and two moons is indicated by the placements of the open circles. For the Earth, advection not associated with plate tectonics (e.g., hotspot volcanism) is insignificant compared to the other two processes

9.2.9.3 Plate Tectonics The mantle is cooled both by advection along the mid-ocean ridges and by heat transfer to the cool, subducting slabs. The Earth is the only known example where this occurs.

9.2.10 Heat Loss Through the Lithosphere of Venus

9.2.10.1 Advection (Volcanoes, Lava Flows) Approximately $200\,\text{km}^3/\text{y}$ of lava would need to flow onto the surface to account for the estimated rate of heat loss from Venus; but cratering ages and visible resurfacing imply $\sim 2\,\text{km}^3/\text{y}$. Therefore, advection is probably not the dominant mechanism.

9.2.10.2 Plate Tectonics An upper limit is given by assuming Aphrodite Terra as an oceanic spreading center. If so, its speed would be centimeters per year; $\sim 0.7\,\text{km}^2/\text{y}$ area would be created and it would produce $\sim 15\%$ of the heat budget. Therefore, most of the heat loss is conductive.

9.2.10.3 Conduction If the heat flux $= 50$–$70\,\text{mW/m}^2$, then the temperature at the base of the lithosphere, $T_{\text{BASE}} = 1650$–$1700\,\text{K}$ ($\sim 100\,\text{K}$ greater than for Earth), the coefficient of thermal conductivity, $\kappa_{\text{LITH}} = 4\,\text{W/(m K)}$, and the thermal lithosphere is about 60–80 km thick or about 50–66% of Earth's.

The conduction may occur primarily at localized hot spots rather than uniformly over the entire lithosphere.

This concludes our brief summary of the surface science of Venus; we now turn to the last of the terrestrial planets, Mars.

9.3 Mars

9.3.1 Visibility and General Properties

Ares in ancient Greece, and Mars from Roman times, this red planet is associated with the god of war. Its brightness at (mean distance) opposition is −2.01, brighter than all but Venus and Jupiter among the planets that can be seen easily at their brightest. It is the lowest of the classical *superior* planets, i.e., those orbiting beyond the orbit of the Sun in geocentric schemes, and the nearest of the planets exterior to Earth in the heliocentric schemes.

In more recent times it was the center of a controversy about extraterrestrial life. Schiaparelli (∼1877) reported the existence of channels (*canali*), visible in telescopic views of Mars in excellent seeing. In the popular English-reporting press, *canali* became "canals," an interpretation eagerly endorsed by Percival Lowell, who claimed 400 canals could be resolved from Earth. Their widths would have had to be ∼50 km or more wide. Lowell thought that the canals channeled water from the melting ice caps to desert-like regions closer to the equator; it was even claimed that a "green wave" swept down from the poles in Martian springtime. Subsequent work has not confirmed these findings (the "green wave" has been shown to be photometrically gray and space probes have ruled out any such long, linear features), although dedicated Martian ground-based observers will readily concede that in very short moments of superlative seeing, there are many more fine details visible than can be remembered long enough to record in a sketch. These experiences are usually chalked up to psychological or physiological causes, but it would be interesting to see speckle or at least adaptive optics work applied to Mars to see if the lineated features have any basis in integrated images from the planet.

Mars' orbit has a relatively large eccentricity, $e = 0.0934$, and an inclination, $i = 1\overset{\circ}{.}850$. Its mean distance is 1.524 au, so its perihelion distance is $a(1 - e) = 1.382$ au, and Mars varies by ±9.3% in its distance from the Sun. On JDN 2450840.5, its ascending node and perihelion longitudes were: $\Omega = 49\overset{\circ}{.}5629$ and $\varpi = 33\overset{\circ}{.}0282$. Its sidereal period is 1.88089 y or $686\overset{d}{.}980$, so that its synodic period is large, in fact the largest of all the planets: $779\overset{d}{.}94$. The studies of Mars by Brahe and Kepler resulted in the discovery of the elliptical nature of the planetary orbits because of the measurable effects of Mars' large eccentricity.

Mars is a rapid rotator, with a sidereal rotation period, $P_{\text{rot}} = 24^{\text{h}}\ 37^{\text{m}}\ 22\overset{s}{.}66 = 1\overset{d}{.}02595675$, compared to Earth's $23^{\text{h}}\ 56^{\text{m}}\ 04\overset{s}{.}10 = 0\overset{d}{.}99726963$. Its rotation axis is tilted by $25\overset{\circ}{.}19$ to the axis of the orbit, so that Mars undergoes seasons as does the Earth, but with nearly twice the lengths of Earth's.

The Martian mean albedo is 0.16. It has a radius of 3397 km, for a mean angular diameter (at mean distance opposition) of $17''.9$. From the semi-major axis of the orbits of Deimos and Phobos, the mass is found from application of Kepler's third law: $M = 6.4191 \times 10^{23}$ kg.

Thus, its compressed mean density is: $<\rho> = 3940\,\text{kg/m}^3$

This is larger than that of the Moon ($3340\,\text{kg/m}^3$) but less than that of all the other terrestrial planets. Its oblateness or flattening is:

$$\varepsilon = 0.006476$$

corresponding to the second spherical harmonic term of the gravitational potential, $J_2 = \{I_z - I_{xy}]/(Mr^2) = 1.964 \times 10^{-3}$. This is the *quadrupole moment*, which measures polar flattening—see Section 5.3. Compare this value to that for the Earth: $J_2 = 1.082626 \times 10^{-3}$. The J_3 term, which indicates asymmetry between N and S hemispheres is also relatively large: $+36 \times 10^{-6}$ (compared to Earth's -2.533×10^{-6}). The Martian ϵ and J_2, J_3 values are the largest among the terrestrial planets.

At the epoch 1998.0, the NCP of Mars was located at $\alpha = 21^{\text{h}}\ 10^{\text{m}}\ 41^{\text{s}}$, $\delta = +52^\circ\ 53'$, $\sim 9^\circ$NW of Deneb (α Cygni). Due to the precession of the Martian equinoxes, it slowly changes.

The mean geometric albedo of Mars is 0.150, and its color indices are: (B-V) = 1.36 and (U-B) = 0.58. These are the reddest colors of all the planets. The general properties of Mars are summarized in Table 9.1.

9.3.2 Global Terrain

The J_3 value indicates that Mars has a large global dichotomy. This is reflected in the physical terrain:

- The southern hemisphere is primarily old, heavily cratered terrain lying 2–3 km above the "mean datum" for Mars.
- The northern hemisphere is primarily younger, lightly cratered terrain lying below the mean datum. The density of cratering is about 2–3 times that on the lunar maria.

The boundary between the two is approximately a great circle making an angle of about 30° with the Martian equator.

9.3.2.1 Volcanoes and Related Features Types of volcanic features seen on Mars include:

Shield volcano:	A volcano with a broad, gently sloping cone; it usually has a shallow caldera at its summit. Examples on Earth: Mauna Loa; Kilauea.
Patera:	A volcano with an even lower profile than a shield volcano, and may be even larger in diameter; it often has a complex caldera at its summit.
Mons:	Literally, "mountain," but on Mars it applies specifically to the largest shield volcanoes.
Tholus:	A smaller, steeper-sided volcano than a shield volcano.
Pyroclastic volcano:	An explosive volcano, producing ejecta. Earth example: Mt Vesuvius.

Specific volcanic features include:

Tharsis Bulge or Ridge. About 4000 km diameter, rising as high as 10 km above the mean datum.

The three large Tharsis volcanoes, *Arsia Mons*, *Pavonis Mons*, and *Ascraeus Mons*, lie about 700 km apart along the "summit ridge" of the Tharsis bulge, and there are several smaller volcanoes on the bulge as well. A *tholus* is a volcanic cone (see Figure 9.26).

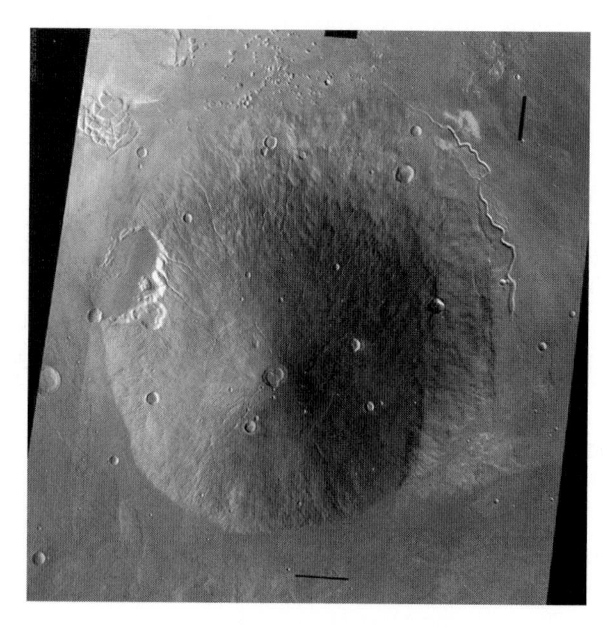

Fig. 9.26. Hecate Tholus, a relatively small volcano; note the many impact craters in the area. A 2001 Mars Odyssey THEMIS image, PIA06827. Credits: NASA/JPL/ASU

Olympus Mons. The largest volcano on Mars, with a base of 600 km diameter and a summit peak of 80 km width and 27 km height. It lies to the NW of Tharsis, and the older *Alba Patera* lies on its northern edge. Olympus Mons appears as a small dark spot on older visual maps. At times this spot appears bright because of clouds, inspiring Schiaparelli to name it Nix Olympica, "Snows of Olympus." Schiaparelli noticed that, during dust storms, it was often one of the few features visible, and concluded correctly that it must be an elevated area. The volcano was renamed Olympus Mons after spacecraft revealed its true identity. Compared with the island of Hawaii, the Earth's largest volcanic shield construct, it is about five times the diameter (~120 km) and three times the height (~9 km, measured from the seafloor bottom). See Figure 9.27.

Alba Patera. A huge ancient shield volcano, in the shape of a shallow dish ("patera"), 1600 km across. Evidently far older than Olympus Mons, it is north of and between Olympus Mons and the Tharsis Ridge volcanoes. Circular cracks around it testify to the great strain on the surrounding material of this volcano, which must have deeply sunk down toward the mantle as it achieved *isostatic equilibrium.*

9.3.2.2 Canyons, Channels and Related Features Some of these features are tectonic[4] in nature (figures are elevation views except where noted as a plan view):

Fig. 9.27. Olympus Mons, at 27 km high, is the highest feature on Mars (dwarfing Mauna Loa and Mt Everest on Earth and Maxwell Montes on Venus). Its caldera is ~80 km across and the entire construct is ~600 km wide. The escarpment itself is 6 km high. Note the clouds around the peak and moving up the flanks. A Mars Global Surveyor mission, Mars Orbital Camera wide field image PIA04737, viewed toward the Martian limb. Release No. MOC2-479, September 10, 2003. Credit: NASA/JPL/Malin Space Science Systems

[4] Tectonic: pertaining to (or caused by or resulting from) structural deformation of the crust.

Fold: deformation of a part of the crust due to cooling and consequent compression.

Anticline: in a sinusoidal deformation, the crest of the fold (the "hill").

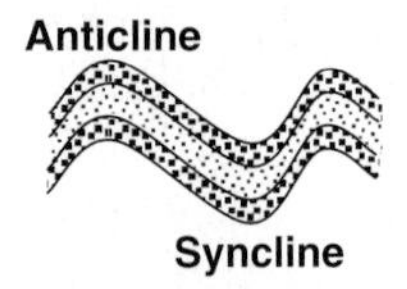

Syncline: in a sinusoidal deformation, the trough of the fold (the "valley").

Monocline: an incomplete fold, characterized by a graduated change from one level to another.

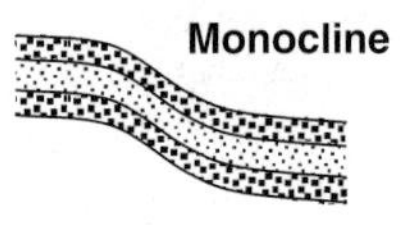

Fault (or rift): a break in the crust (instead of folding).

Normal fault: vertical displacements, in opposite directions, of adjacent pieces of crust.

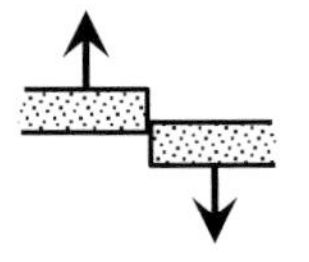

Thrust fault (or reverse fault): one piece of crust rides up over an adjacent piece of crust, which slides under.

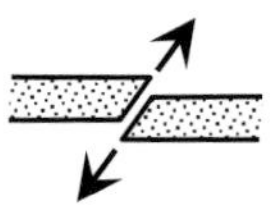

Strike–slip fault: horizontal displacements, in opposite directions, of adjacent pieces of crust.

Graben: a piece of crust that has dropped between two normal faults.

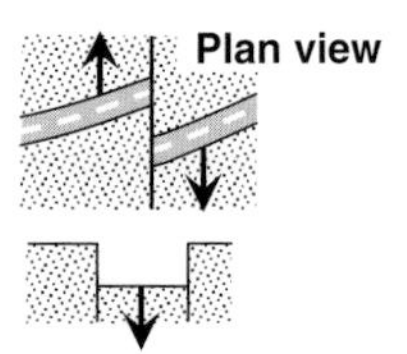

Tectonic and/or erosional features:

Canyon systems. The largest of these is Valles Marineris (45°W to 90°W, near the equator). This is an extensional rift system about 4000 km long, extending eastward from the middle of the Tharsis bulge. The valley system is up to 500 km wide and 7 km deep (see Figure 9.28). It includes, among many, the following major canyons:

Ius Chasma
Tithonius Chasma (75 km wide, 4 km deep)
Coprates Chasma

Outflow Channels. These large-scale features show the presence of immature tributaries and are widespread. They may be as much as hundreds of kilometers long, starting and stopping abruptly. See Figures 9.29 and 9.30. Some characteristics (not always present) are:

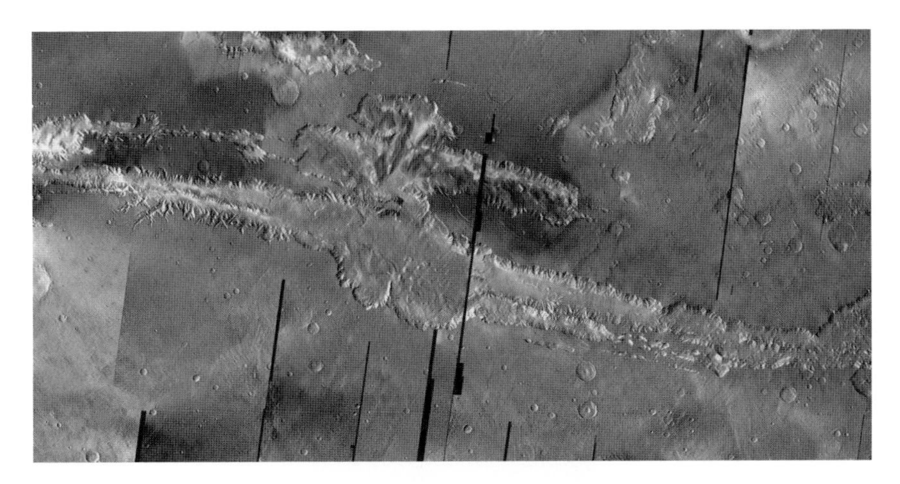

Fig. 9.28. The western portion of Valles Marineris, the major rift system on Mars, extending 4000 km east–west and as wide as 500 km in places. Note the chains of collapse pits outside the canyon's south side. The loss of subsurface ice may contribute to the growth of the canyon system. A 2001 Mars Odyssey spacecraft mosaic, NASA/JPL/Arizona State University image PIA06926

- Tear-drop shaped islands and bars (as, for example, in the Ares Vallis)
- Meanders (one particularly extensive one closely resembles the Red River on Earth)
- Sometimes braided patterns in the region, resembling post-flood debris deposition
- Sometimes found in chaotic terrain

An example is *Maja Vallis* in *Lunae Planum*

Fossae[5] Extensional fractures (grabens), usually occurring as fracture systems.

Dendritic channels, gullies or arroyos. These resemble drainage channels, and are found on crater walls and elsewhere. They may be as much as 10s of kms long. On some crater walls, they appear to emerge from a particular layer; these regions do not appear very ancient, for the most part. Outbursts of both CO_2 and H_2O have been suggested as possible causes. The features are found more frequently at higher latitude sites, which tend to be well below the freezing point of water, thus favoring landslides created by outbursting

[5] *Fossa* is the Latin word for a ditch or trench. *Fossae* is the plural. See Figures 9.31 and 9.32 for examples on Mars.

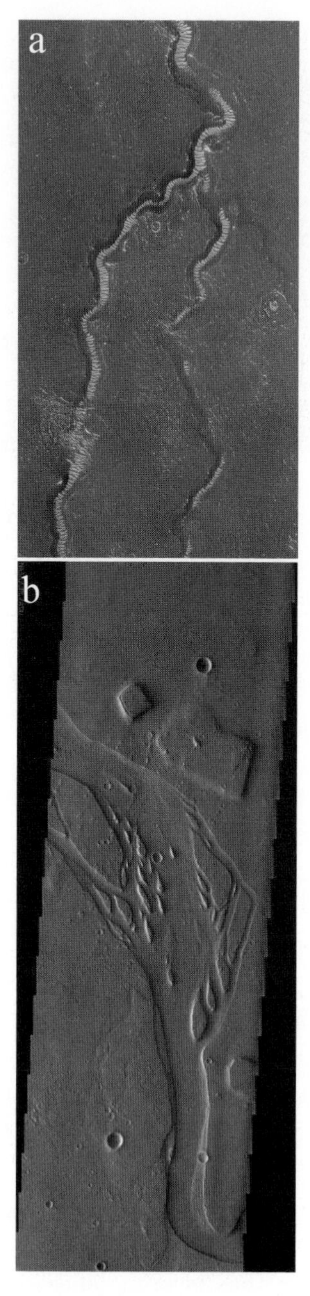

Fig. 9.29. **a** Apsus Valley channels, with meanders and sand dunes. MGS Mars Orbital Camera image, PIA05992. Credits: NASA/JPL/Malin space Science System. **b** SE of Elysium Mons, with tear-drop islands. 2001 Mars Odyssey THEMIS images, PIA04586. Credits: NASA/JPL/Arizona State University (ASU)

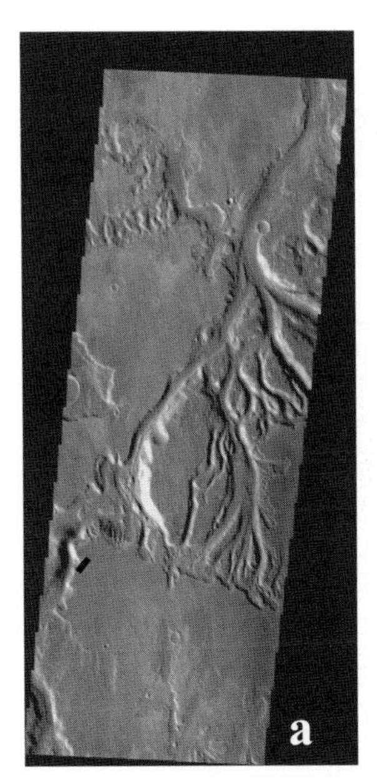

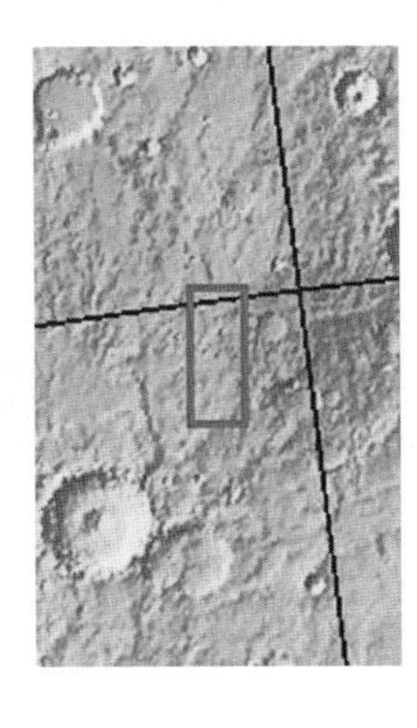

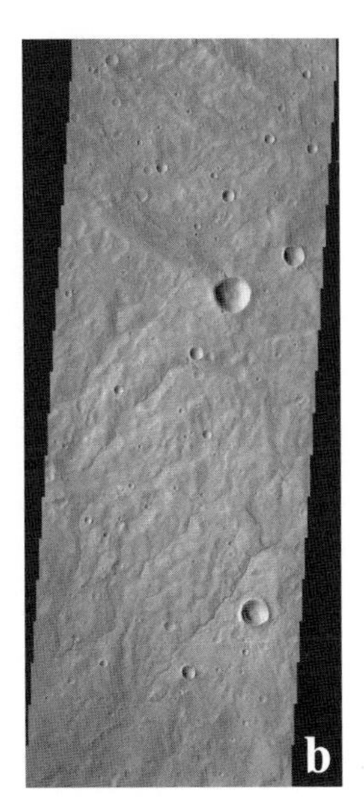

Fig. 9.30. **a** The Sabis Vallis. Note the tributaries. Image PIA03663. **b** Channels joining to form Sabis Vallis. Image PIA03664, the context for which is shown in the middle. Credits: NASA/JPL/ASU. 2001 Mars Odyssey THEMIS VIS images produced by Arizona State University (ASU)

pockets of CO_2. On the other hand, the ice content of the surface layers is expected to be more deeply buried nearer the equatorial regions, where a higher rate of sublimation would long ago have exhausted the near-surface ice.

9.3.2.3 Other types of terrain *Fretted terrain.* Flat lowlands bordered by steep cliffs and filled with elevated plateaus (mesas). They are due to erosional recession of the highlands.

Chaotic terrain. These are depressed areas characterized by jumbled slabs and blocks. They are likely due to subsurface withdrawal. Examples: *Hydaspis chaos*, ~100 km wide; *Aureum chaos* (see Figure 9.33).

Layered terrain. The earliest known and best examples are near the poles. They are stacked layers, each ~30 m thickness, due mainly to aeolian deposits of sediments. Global surveyor images have also shown layered terrain in basins and craters; these tend to be more frequent at lower latitude sites.

Etched plains. These are located in polar regions, and are pits and hollows possibly due to saltating particles and sublimation (primarily of CO_2 ice).

Fig. 9.31. As seen from the Odyssey 2001 spacecraft, a graben, part of Cerberus Fossae in a volcanic region of Mars. Note the debris at the bottom of the feature where the surface material has collapsed between two faults. A 2001 Mars Odyssey THEMIS VIS instrument image PIA06842. Credits: NASA/JPL/ASU

Dust deposits. Barchan dunes (crescent-shaped with apex facing the wind), *transverse dunes* (rows of overlapping crescents, apices all facing the wind), and *longitudinal dunes* (more or less parallel, linear, carved by shifts in wind direction) are seen in many areas, including floors of wide craters. Sometimes dusty regions can be identified by contrast to dark streaks downwind of isolated dome or mesa-like structures, which serve to block aeolian (wind-borne) deposits. Figure 9.34 illustrates a variety of dunes.

Impact basins. There are 19 known impact basins over 250 km diameter; the five largest are:

Hellas	2000 km dia
Isidis	1900 km dia
Argyre	1200 km dia
South Polar	850 km dia
Chryse	800 km dia

For comparison, the Moon's Mare Imbrium has a diameter of 1300 km.

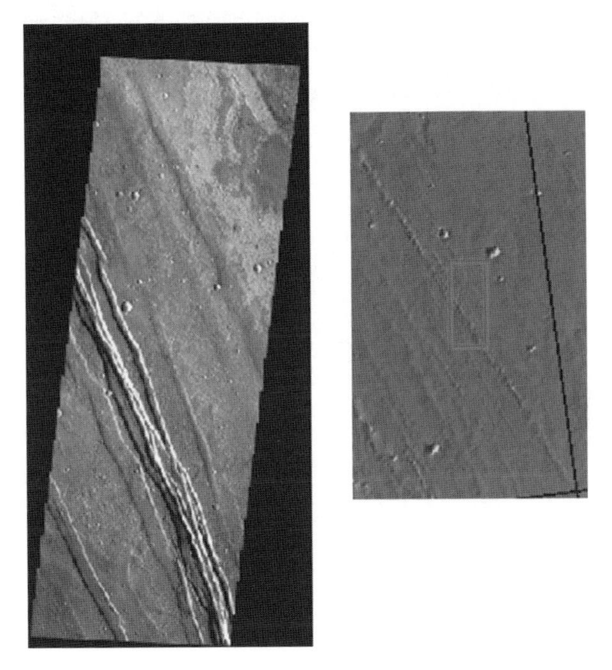

Fig. 9.32. Multiple parallel faults at 23°S and 259°E in the region between Syria Planum and Claritas Rupes on Mars. A 2001 Mars Odyssey THEMIS VIS instrument image, NASA/JPL/ASU image PIA02297, and context image, produced by Arizona State University, Tempe

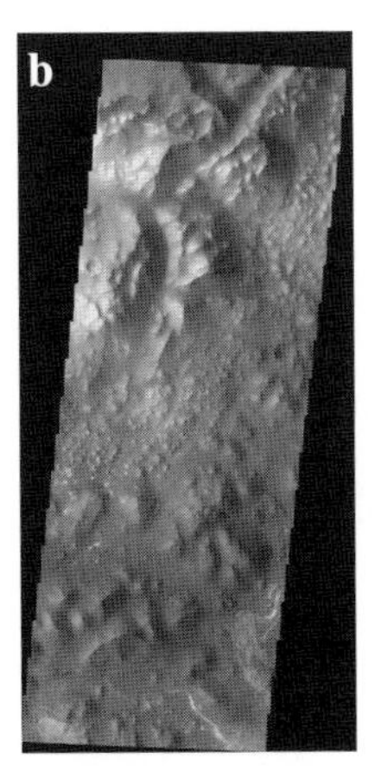

Fig. 9.33. Aureum Chaos, 3.6°S, 333°E. **a** Context image, showing the highlands; **b** blocks and deposition as seen from the Mars Odyssey 2001 spacecraft. THEMIS VIS instrument images, PIA02196. Credits: NASA/JPL/ASU

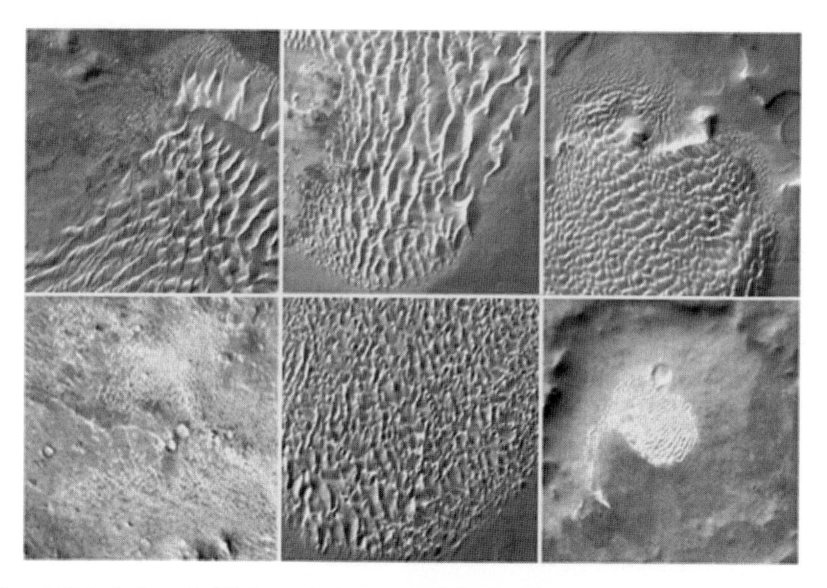

Fig. 9.34. Infrared (12.6 μm) images of dunes from various areas of Mars. NASA/JPL 2001 Mars Odyssey THEMIS VIS image PIA03740 produced by Arizona State University

9.3.3 Evidence of Climate Change on the Martian Surface

In the late nineteenth and earlier twentieth century, speculation was rife that Mars was a dynamic place where water was scarce but skillfully managed by an engineering civilization. Following the first spacecraft views provided by the early Mariner missions, Mars was viewed as a dead world, one characterized by little atmosphere and waterless deserts that were dominated by enormous numbers of impact craters. Since then, the view has gradually shifted to that of a planet currently caught in a deep ice age, but which once possessed a much wetter and warmer climate.

The proximity of the huge tectonic rift valley complex in the vicinity of the enormous volcanic constructs suggests a strong connection. It seems likely that tectonic forces caused strong deformation of the crust both in and around Tharsis Ridge. Since the construct required a very large reservoir of magma, the massive outpouring of this material had to cause a withdrawal elsewhere; this may have led to the development of the fault valleys. The presence of linear series of collapse pits confirms a tectonic origin for the Valles Marineris, but subsequent erosion has clearly occurred, as indicated by the branched gullies on the canyon rims (Figure 9.28). The volcanism involved extensive outgassing and since water vapour and carbon dioxide are greenhouse gases, sufficient amounts may have been emitted to alter the

Martian climate, permitting rain and surface water to produce the erosional effects so clearly visible today. The estimated date of the volcanism is $\sim 10^9$ years ago.

The Global Surveyor has produced evidence of changing CO_2 coverage of Mars' south polar cap. The evidence as revealed in images 3 years apart, however, is probably insufficient to justify the conclusion that Mars' climate is changing rapidly, because the phenomenon has not been observed at high spatial resolution (25–100 m/pixel) for a sufficient length of time to gauge the normal range of fluctuation of the CO_2 coverage.

Mars satellite instruments, most recently those on the Global Surveyor, have revealed images suggesting the presence of large if shallow ocean basins, regions of what appear to be dry water courses, and apparent water outflow sources that seem to imply present processes at work. One such low-altitude region covers a large area around the North Pole. The interpretations are moot, however; what is safest to say is that both source and mechanism for the recent-looking channels are unknown at present. However, the outflow from the great canyons clearly tends toward the Chryse Plain, the site of one of the two 1976 Viking Landers. The triggering of floods is usually explained by volcanism and impacts. The debris apron, sometimes cited as "lobate ejecta," in the vicinity of many impact craters strongly suggest the existence of great mudslides of melted permafrost (Figure 9.35). In addition to catastrophic floods, however, there is now ample evidence that flowing water must have been sustained over lengthy periods of time.

The series of observations obtained starting in 2003 by the Spirit and Opportunity rovers on opposite sides of the Martian surface have clinched the argument that both flowing and standing water has been present on Mars in the past. The durations of these aqueous events are unknown at present. Salts, and "blueberry" concretions (Figures 9.36 and 9.37) that needed to be formed in a water environment, have been found, and so have *festoons*, a kind of rippling pattern seen in the rock. Widespread layering indicates clearly that deposition must have been episodic, whether aolean or water-borne. Layering is also cited as evidence for a widespread and long-lasting water environment, but polar strata can be understood as the seasonal interplay of volatile ices (on Mars, water vapor and carbon dioxide) as they sublime and recondense, and the widespread and voluminous dust that coats it during the seasonal global sandstorms. Similarly, the layering seen in Canyon walls could have been laid down in a liquid medium (Figures 9.38 and 9.39), but aside from that in the strata studied by the Rovers, the large-scaled rock layers could well be products of aeolean deposition.

Arguments for evidence of past Martian life forms in the Antarctic rock, ALH 84001, are more controversial (see below and Milone & Wilson, 2008, Chapter 15.3), but the age of Mars exploration is only beginning, and many exciting results can be anticipated over the next decade. Other NASA and

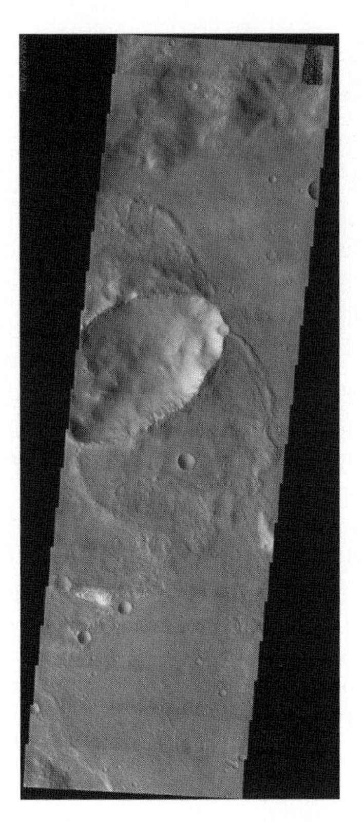

Fig. 9.35. "Lobate ejecta" patterns can be seen around this large but non-circular crater located at 24°.6 S, 41°E; its "butterfly" pattern is likely due to an oblique impact. Taken with THEMIS aboard the Odyssey 2001 spacecraft. A NASA/JPL/Arizona State University image

ESA missions will be sent to Mars and will be searching for further evidence of the existence, past or present, of life.

The surface of Mars is becoming better known, but the interior is almost completely unknown. The evidence of great volumes of lava in many volcanoes and constructs indicates that at least at one time, Mars was a very active planet. The presence of such huge constructs as Olympus Mons and the Tharsis Ridge indicates that Mars probably does not have plate tectonics, but certainly has undergone extensive tectonic events, with stationary mantle plumes. At present there is no evidence of active volcanism, but the appearance of many possibly water-fed gullies suggests that some kind of interior warming may act from time to time to sublime or melt the subsurface ice, which can then explode through porous strata and cause the observed features. From thermal profile models, Mars is estimated to have a heat flux

Fig. 9.36. From the Mars Exploration Rover (MER) *Opportunity* in Meridiani Planum: "blueberries," concretions formed in porous rock, and "festoons" (sinuous markings), thought to arise from ripples in an aqueous environment. Credits: NASA/JPL-Caltech/Cornell. PIA03279

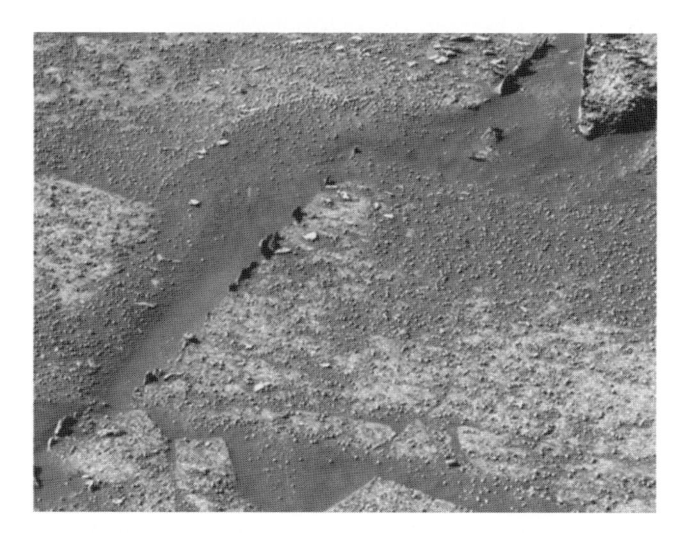

Fig. 9.37. The Mars Exploration Rover *Opportunity* found evidence of a water environment in Endurance Crater in the form of flow channels and "blueberries," shown here. Credits: NASA/JPL/Cornell. Image PIA06692 (See also Plate 6)

Fig. 9.38. A sub-frame of an image from the Mars Reconnaisance Orbiter High Resolution Imaging Science Experiment (HiRISE), in false color to delineate reflectance from different materials. This view of a scarp of Chasma Boreale, near the north pole of Mars shows layered deposits overlying darker material. Note the interweaving of bright, ice-laden layers with darker sand layers. NASA/JPL-CalTech/University of Arizona image PIA09097 (See also Plate 7)

output of 30 mW/m^3 (Carr 1999). This helps to explain why gullies tend to be seen emerging from particular strata in features located at higher latitude sites, where the subsurface ice is found closer to the surface, than from sites near the equator. Figure 9.40 demonstrates gullies at several sites.

Fig. 9.39. Layering at a precipice at Terby Crater, north of Hellas, near 27.6S, 286W. Mars Global Surveyor Mars Orbiter Camera image PIA04582. Credits: NASA/JPL/Malin Space Science Systems image

We conclude this chapter with a brief discussion of the evidence that traces of organic life may have been detected in a rock from Mars.

9.3.4 The Evidence for Past Life on Mars

The announcement at a NASA press conference in 1996 of the possible discovery of evidence for past life on Mars was electrifying. McKay et al. (1996) cited mineralogical and morphological evidence for their conclusions about a meteorite from Mars recovered from Antarctica, ALH 84001 (described in Section 15.3). Current evidence includes the presence of magnetite, possibly biogenically-produced, and polycyclic aromatic hydrocarbons (PAHS). However, a number of objections have been raised: possible non-biogenic origin for the 10–200 nm-scale carbonate (Treiman et al., 2002; Treiman 2003) and magnetite (Golden et al., 2001) structures; the segmented

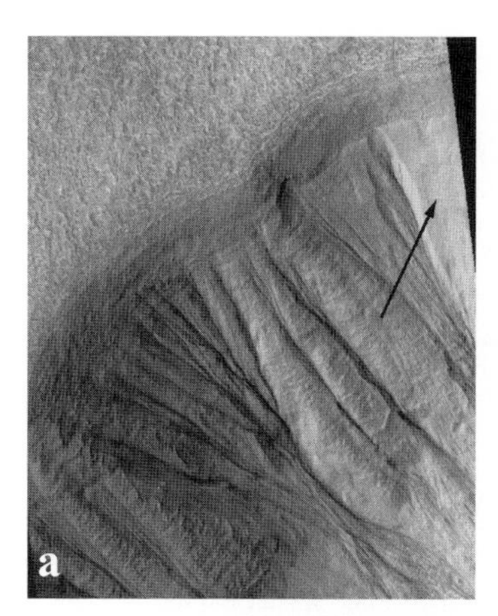

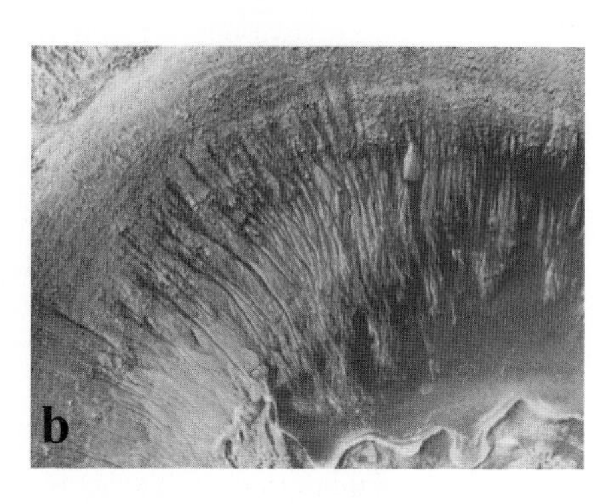

Fig. 9.40. Mid-latitude crater wall gullies. Mars Global Surveyor (MGS) Mars Orbiter camera images. **a** Image M09-2875 (PIA04409) of a site 33°S and 93°E showing a 2.8 × 4.5 km view. The arrow indicates a possible snowpack source for the outflows. Credits: NASA/JPL/Malin Space Science Systems/Philip Christensen. **b** Image MOC2-242A PIA02824/PIA01039. The north wall of a 7-km diameter crater within the 287-km wide Newton crater at 41°S, 160°W. The gullies are characterized by alcoves due to subsidence, at the source, the outflow channels (of water or CO_2), and the apron of debris at the base. Similar characteristics are seen on such features on Earth. Credits: NASA/JPL/Malin Space Science Systems

forms (Figure 9.41) were but artifacts of the imaging process (Bradley et al., 1997); such small organisms could not contain the structures needed to metabolize; and the terrestrial contamination of the material in the sense that a significant fraction of the carbon present is ^{14}C, dispersed from atomic bomb tests of the 1950s (Jull et al., 1997); and other, mineralogical arguments. Although vigorously defended (e.g., by McKay et al., 1997; McKay et al., 2002; Thomas-Keprta et al. 2002), the issue remains

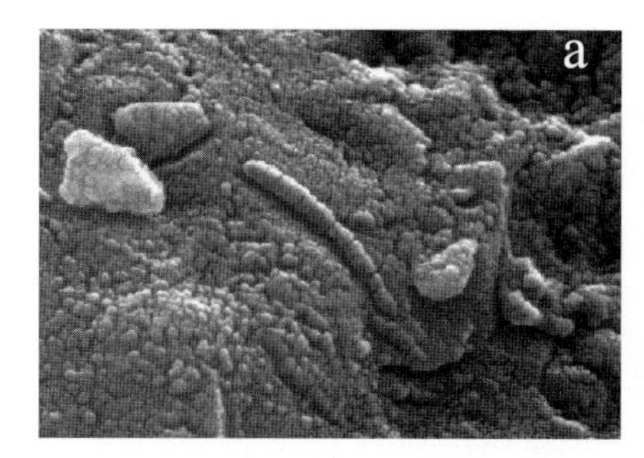

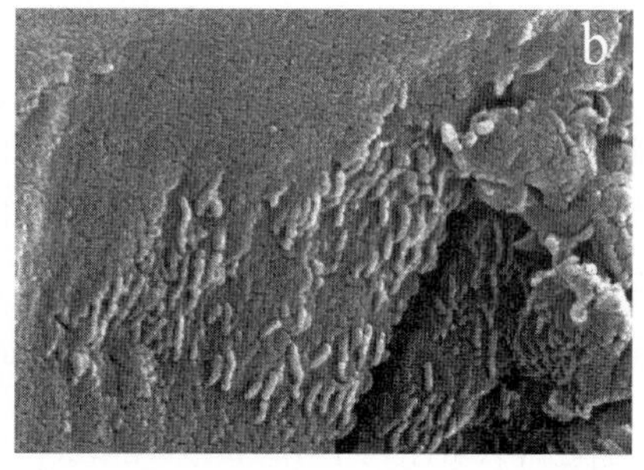

Fig. 9.41. Possible morphological evidence for the presence of microfossils (or parts or fragments thereof) in segmented carbonate deposits in the Martian meteorite ALH 84001. NASA/AMLAMP images **a** PIA00288 and **b** PIA00284

controversial, with many, perhaps most, planetary scientists rejecting the claim.

This concludes our examination of the surfaces and interiors of the terrestrial planets. The atmospheres of planets will be considered in Chapter 10, and their ionospheres and magnetospheres in Chapter 11, of Milone & Wilson, 2008.

References

Abell, G. 1982. *Exploration of the Universe* 4th ed. (Philadelphia, New York...: Saunders College Publishing; and earlier editions by Holt, Rinehart and Winston)

Benz, W., Stattery, W.L., & Cameron, A.G.W. 1998. "Collisional Stripping of Mercury's Mantle," *Icarus* **74**, 516–528.

Bradley, J. P., Harvey, R. P., and McSweem, Jr., H. Y. 1997. "No 'Nanofossils' in Martian Meteorite," *Nature*, **390**, 454.

Carr, M. H. 1999. "Mars: Surface and Interior," in *Encyclopedia of the Solar System*, eds P. R. Weissman, L.-A. McFadden, and T. V. Johnson (San Diego, CA: Academic Press), pp. 291–308.

Consolmagno, G. J. and Schaefer, M. W. 1994. *Worlds Apart: A Textbook in the Planetary Sciences* (Englewood Cliffs, NJ: Prentice Hall).

Dicke, R. H. and Goldenberg, H. M. 1967. "Solar Oblateness and General Relativity," *Physical Review Letters*, **18**, 313–316.

Fegley Jr., B., 2005. "Venus," Chapter 21, in *Meteorites Comets, and Planets*, ed. A. M. Davis. Volume I of *Treatise of Geochemistry*, eds H. D. Holland and K. K. Turekian (Oxford: Elsevier-Pergamon).

Fegley Jr., B., Klinglehöfer, G., Lodders, K., and Widemann, T. 1997. "Geochemistry of Surface–Atmosphere Interactions on Venus," in *Venus II: Geology, Geophysics, Atmosphere, and Solar Wind Environment*, eds S. W. Bougher, D. M. Hunten, and R. J. Phillips (Tucson, AZ: University of Arizona Press), pp. 591–636.

Golden, D. C., Ming, D. W., Schwandt, C. S., Lauer, H. V., Socki, R. A., Morris, R. V., Lofgren, G. E., and McKay, G. A. 2001. "A Simple Inorganic Process for Formation of Carbonates, Magnetites, and Sulfides in Martian Meteorite ALH84001," *American Mineralogist*, **86**, 370–375.

Head, J. W. and Basilevsky, A. T. 1999. "Venus: Surface and Interior," in *Encyclopedia of the Solar System*, eds P. R. Weissman, L.-A. McFadden, and T. V. Johnson (San Diego, CA: Academic Press), pp. 161–189.

Jull, A. J. T., Courtney, C., Jeffrey, D. A., and Beck, J. W. 1997. "Isotopic Evidence for a Terrestrial Source of Organic Compounds Found in Martian Meteorites Allan Hills84001 and Elephant Moraine 79001," *Science,* **279**, 366–369.

Kelley, D. H. and Milone, E. F. 2005. *Exploring Ancient Skies: An Encyclopedic Survey of Archaeoastronomy.* (New York: Springer)

Kerr, R. A. 1991. "Magellan: No Venusian Plate Tectonics Seen," *Science,* **252**, 213.

Lewis, J. S. 1995. *Physics and Chemistry of the Solar System* (San Diego, CA: Academic Press)

Lissauer, J. L. 1993. "Planet Formation," *Annual Review of Astronomy and Astrophysics*, **31**, 129–174.

Margot, J. L., Peale, S. J., Jurgens, R. F., Slade, M. A., and Holin, I. V. 2007. "Large Longitude Libration of Mercury Reveals a Molten Core," *Science*, **316**, 710–714.

McKay, D. S., Clemett, S.J., Gibson, E. K., Jr., Thomas-Keprta, K., and Wentworth, S. J. 2002. "Are Carbonate Globules, Magnetites, and PAHs in ALH84001 Really Terrestrial Contaminants?" *Lunar and Planetary Society*, **XXXIII**, Pdf. 1943.

McKay, D. S., Gibson, Jr., E. K., Thomas-Keprta, K. L., Vali, H., Romanek, C. S., Clemett, S. J., Chillier, X. D. F., Maechling, C. R., and Zare, R. N. 1996. "Search for Past Life on Mars: Possible Relic Biogenic Activity in Martian Meteorite ALH84001," *Science*, **273**, 924–930.

McKay, D. S., Gibson, Jr., E., Thomas-Keprta, K. 1997. "Reply to: No 'Nanofossils' in Martain Meteorites," *Nature,* **390,** 465–466.

Milone, E. F., and Wilson, W. J. F. 2008. *Solar System Astrophysics: Planetary Atmospheres and the Outer Solar System.* (New York: Springer)

Misner, C. W., Thorne, K. S., and Wheeler, J. A. 1973. *Gravitation* (San Francisco, CA: W. H. Freeman).

Ross, F. E. 1928. "Photographs of Venus," *Astrophysical Journal*, **68**, 57–92.

Sandwell, D. T. and Schubert, G. 1992. "Evidence for Retrograde Lithospheric Subduction on Venus," *Science*, 257, 766–770.

Strom, R. G. and Sprague, A. L. 2003. *Exploring Mercury: The Iron Planet* (Berlin: Springer-Verlag; Chichester, UK: Praxis).

Taylor, S. R. 1992. *Solar System Evolution: A New Perspective* (Cambridge: University Press).

Thomas-Keprta, K. L., Clemett, S. J., Bazylinski, D. A., Kirschvink, J. L., McKay, D. S., Wentworth, S. J., Vali, H., Gibson, Jr., E. K., and Romanek, C. S. 2002. "Magnetofossils from Ancient Mars: A Robust Biosignature in the Martian Meteorite ALH84001," *Applied and Environmental Microbiology*, **68**, No. 8, 3663–3672.

Treiman, A. H. 2003. "Submicron Magnetite Grains and Carbon Compounds in Martian Meteorite ALH84001: Inorganic, Abiotic Formation by Shock and Thermal Metamorphism," *Astrobiology*, **3**, 369–392.

Treiman, A. H., Amundsen, H. E. F., Blake, D. F., and Bunch, T. 2002. "Hydrothermal Origin for Carbonate Globules in Martian Meteorite ALH84001: A Terrestrial Analogue from Spitzbergen (Norway)," *Earth and Planetary Science Letters*, **204**, 323–332.

Wetherill, G. W. 1988. in *Mercury*, ed. F. Vilas et al. (Tucson, AZ: University of Arizona Press), pp. 670–691.

Challenges

[9.1] Calculate (a) the escape velocity of an atom from the surface of Mercury and (b) the rms speeds of atoms of S and Fe for a 15,000 K surface on the planet Mercury. (c) Comment on the retention of these atoms for the rapid accretion scenario described in Section 9.1.

[9.2] Compute the impact speed and specific impact energy for an asteroid colliding with each of the terrestrial planets and the Earth's moon. Assume the asteroid to have an orbit with the same semi-major axis as the orbit of the planet.

[9.3] Ignore atmospheric effects for the situation in [9.2] and comment on the size of the craters one would expect for each body for the same mass of impactor of a stony meteorite (say $\rho = 3500\,\mathrm{kg/m^3}$). Is it reasonable to suppose one should ignore atmospheric effects?

[9.4] Estimate the mass of impactor required to create the Hellas basin on Mars. Show all reasoning.

[9.5] Compare the observed and global equilibrium temperature of Venus. Is it reasonable to ignore internal heat sources on this planet? If the only source of heat on Venus were internal, compute its equilibrium temperature.

Index

Printed in the United States of America